Studies in Fuzziness and Soft Computing 281

Editor-in-Chief

Prof. Janusz Kacprzyk
Systems Research Institute
Polish Academy of Sciences
ul. Newelska 6
01-447 Warsaw
Poland
E-mail: kacprzyk@ibspan.waw.pl

For further volumes:
http://www.springer.com/series/2941

Zaiwu Gong, Yi Lin, and Tianxiang Yao

Uncertain Fuzzy Preference Relations and Their Applications

Authors
Dr. Zaiwu Gong
College of Economics and Management
Nanjing University of Information Science
and Technology
Nanjing
PR China

Dr. Tianxiang Yao
College of Economics and Management
Nanjing University of Information Science
and Technology
Nanjing
PR China

Dr. Yi Lin
Department of Mathematics
Slippery Rock University
USA

ISSN 1434-9922 e-ISSN 1860-0808
ISBN 978-3-642-28447-2 e-ISBN 978-3-642-28448-9
DOI 10.1007/978-3-642-28448-9
Springer Heidelberg New York Dordrecht London

Library of Congress Control Number: 2012933080

© Springer-Verlag Berlin Heidelberg 2013
This work is subject to copyright. All rights are reserved by the Publisher, whether the whole or part of the material is concerned, specifically the rights of translation, reprinting, reuse of illustrations, recitation, broadcasting, reproduction on microfilms or in any other physical way, and transmission or information storage and retrieval, electronic adaptation, computer software, or by similar or dissimilar methodology now known or hereafter developed. Exempted from this legal reservation are brief excerpts in connection with reviews or scholarly analysis or material supplied specifically for the purpose of being entered and executed on a computer system, for exclusive use by the purchaser of the work. Duplication of this publication or parts thereof is permitted only under the provisions of the Copyright Law of the Publisher's location, in its current version, and permission for use must always be obtained from Springer. Permissions for use may be obtained through RightsLink at the Copyright Clearance Center. Violations are liable to prosecution under the respective Copyright Law.
The use of general descriptive names, registered names, trademarks, service marks, etc. in this publication does not imply, even in the absence of a specific statement, that such names are exempt from the relevant protective laws and regulations and therefore free for general use.
While the advice and information in this book are believed to be true and accurate at the date of publication, neither the authors nor the editors nor the publisher can accept any legal responsibility for any errors or omissions that may be made. The publisher makes no warranty, express or implied, with respect to the material contained herein.

Printed on acid-free paper

Springer is part of Springer Science+Business Media (www.springer.com)

Preface

In any real-life decision making situation, the judgment matrices or preference relations that reflect the experience and the judgment of the experts (or decision makers) represent an effective way to settle problems that seem to be nonstructural and mix both qualitative and quantitative analyses. However, it is difficult for the experts to present precise, consistent judgment, and achieve the optimal decision making because of their limited knowledge, bounded rationality in human thinking and reasoning. Hammond (1996) believes that intuitional judgment is more accurate than quantitatively precise judgment due to how human mind works. Technically, intuitional judgments are represented and characterized by different kinds of uncertain fuzzy sets. In terms of interval fuzzy sets, they are expressed by using continuous intervals, triangular fuzzy numbers, or trapezoidal fuzzy numbers. In terms of fuzzy logic and soft computing, they are shown by applying linguistic fuzzy numbers. In terms of intuitionistic fuzzy sets, they are represented by utilizing intuitioinstic fuzzy numbers.

In this book, from the view point of the basic theory of fuzzy sets and its generalizations, we discuss various theories of uncertain preference relations, including consistency theories, aggregation theories, modeling methods of selecting satisfactory decision alternatives, and the corresponding intelligent algorithms. From the view point of applications, we discuss the feasibility and rationality of these theoretical studies by making use of actual numerical cases. This book is the fruition of authors' seven years (from 2004 to 2011) research.

The book consists of eight chapters. The main relationship among all these chapters is the development of fuzzy sets and its generalized theory. The main structural framework of each chapter includes consistency theories, aggregation theories, modeling methods of selecting satisfactory decision alternatives, and the corresponding intelligent algorithms. In particular, the first chapter introduces the connotation of uncertainty, the research range of uncertain judgment knowledge, and the research contents of uncertain preference relations. Chapter 2 deals with fuzzy preference relations and their related theories. It is in this chapter that we introduce such basic concepts as consistency, priority of a reciprocal preference relation, and fuzzy preference relations. Additionally, we present a method of group decision making based on fuzzy preference relations of incomplete information.

Chapter 3 studies preference relations of interval fuzzy numbers and their related theories, such as the consistency theories and priority models of preference relations of interval fuzzy numbers. As a consequence, a method of group decision making based on incomplete preference relations of interval fuzzy numbers is developed. Chapter 4 considers preference relations of triangular fuzzy numbers and their relevant theories, where consistency theories and priority models of preference relations of triangular fuzzy numbers are established. At the conclusion, we put forward a method of group decision making using incomplete preference relations of triangular fuzzy numbers. Chapter 5 investigates preference relations of 2-tuple linguistic terms, where after summarizing the development of linguistic preference relations, we build consistency theories and group decision making theories on the basis of preference relations of 2-tuple linguistic terms.

Chapter 6 is devoted to the study of preference relations of trapezoidal fuzzy numbers. In this chapter, we discuss possible transformations between preferences of trapezoidal fuzzy numbers and those of 2-tuple linguistic terms. Then after summarizing the development of linguistic preference relations, we look at the aggregation, priority model and build consistency theories and group decision making theories based on 2-tuple linguistic preference relations and preference relations of trapezoidal fuzzy numbers. Chapter 7 deals with group decision making by employing different fuzzy preference relations, where we propose methods of aggregation and group decision making, and applications of multiple preference relations.

In the first part of Chapter 8, we develop consistency and priority theories, and group decision making by using incomplete preference relations of intuitionistic fuzzy values. In the second part of Chapter 8, we construct theories for analyzing group consensus and optimization models for group decision making on the basis of intuitionistic fuzzy preference relations. Then, this book is concluded with some final words.

This work has been jointly supported by National Natural Science Foundation of China (No. 70901043 and No. 71171115), funded projects of Ministry of Education of PR China for young scholars who are devoted to the Research of Humanities and Social Sciences (09YJC630130, 09YJC630129), Qing Lan Project (No. 0911), Philosophical and Social Science Foundation of Higher Education of Jiangsu Province of China under the grant (09SJB630043), Natural Science Foundation of Higher Education of Jiangsu Province of China under the grant (08KJD630002), and Research and Practice of Education and Teaching Reform for Graduate Learning of Jiangsu Province (JGKT10034). At this junction, we also extend our heartfelt thanks for the constructive discussions and various supports of Dr. Sifeng Liu and Dr. Lianshui Li.

It is our hope that this book will capture the academic attention of many colleagues from around the world. If you, the reader of this book, like to communicate with any of us, please drop us a line at zwgong26@163.com for Dr. Zaiwu Gong, jeffrey.forrest@sru.edu for Dr. Yi Lin (also known as Jeffrey Yi-Lin Forrest), or ytxnj@163.com for Dr. Tianxiang Yao.

October 30, 2011

Zaiwu Gong
Yi Lin
Tianxiang Yao

About the Authors

Dr. Zaiwu Gong received his PhD degree in management science and engineering from Nanjing University of Aeronautics and Astronautics (China). He is an associate dean of College of Economics and Management, Nanjing University of Information Science and Technology. He is a director of Professional Committee of Grey System of Chinese Society of Optimization, Overall Planning and Economic Mathematics. He was a visiting scholar of Pennsylvania State Systems of Higher Education (Slippery Rock campus) during January – June 2011. He has been a reviewer of several international journals, such as Information Science, International Journal of Approximate Reasoning, and International Journal of Computational Intelligence Sciences.

He is responsible of National Natural Science Foundation of China, Humanities and Social Sciences Foundation of Ministry of Education of China and Philosophical, Social Science Foundation of Higher Education of Jiangsu Province of China and Natural Science Foundation of Higher Education of Jiangsu Province of China, Sciences Foundation of Nanjing University of Information Science and Technology, and Research and Practice of Education and Teaching Reform for Graduate of Jiansu Province. And he has also participated in the operations of over 20 other foundations.

He has published over 50 research papers. His research interests are mainly in the areas of group decision making, fuzzy multi-attribute decision making, fuzzy mathematics, meteorological engineering & management and hazard evaluation. He was recognized as an eminent teacher of Qinglan Project. He was awarded an award of distinguished doctoral dissertation of Nanjing University of Aeronautics and Astronautics in 2009.

Dr. Yi Lin, also known as Jeffrey Yi-Lin Forrest, holds all his educational degrees (BS, MS, and PhD) in pure mathematics from Northwestern University (China) and Auburn University (USA) and had one year of post-doctoral experience in statistics at Carnegie Mellon University (USA). Currently, he is a guest or specially appointed professor in economics, finance, systems science, and mathematics at several major universities in China, including Huazhong University of Science and Technology, National University of Defense Technology, Nanjing University of Aeronautics and Astronautics, and a tenured professor of mathematics at the Pennsylvania State System of Higher Education (Slippery Rock campus). Since 1993, he has been serving as the president of the International Institute for General Systems Studies, Inc. Along with various professional endeavors organized by him, Dr. Lin has had the honor to mobilize scholars from over 80 countries representing more than 50 different scientific disciplines.

Over the years, he has and had served on the editorial boards of 11 professional journals, including Kybernetes: The International Journal of Systems, Cybernetics and Management Science, Journal of Systems Science and Complexity, International Journal of General Systems, and Advances in Systems Science and Applications. And, he is a co-editor of the book series entitled "Systems Evaluation, Prediction and Decision-Making," published Taylor and Francis since 2008.

Some of Dr. Lin's research was funded by the United Nations, the State of Pennsylvania, the National Science Foundation of China, and the German National Research Center for Information Architecture and Software Technology.

By the end of 2008, he had published nearly 300 research papers and over 30 monographs and edited special topic volumes by such prestigious publishers as Springer, Wiley, World Scientific, Kluwer Academic (now part of Springer), Academic Press (now part of Springer), and others. Throughout his career, Dr. Yi Lin's scientific achievements have been recognized by various professional organizations and academic publishers. In 2001, he was inducted into the honorary fellowship of the World Organization of Systems and Cybernetics.

Professor Yi Lin's professional career started in 1984 when his first paper was published. His research interests are mainly in the area of systems research and applications in a wide-ranging number of disciplines of the traditional science, such as mathematical modeling, foundations of mathematics, data analysis, theory and methods of predictions of disastrous natural events, economics and finance, management science, philosophy of science, etc.

Dr. Tianxiang Yao received his PhD degree in management science and engineering from Nanjing University of Aeronautics and Astronautics (China). He is a lecturer at College of Economics and Management, Nanjing University of Information Science and Technology. He is a director of Professional Committee of Grey System of Chinese Society of Optimization, Overall Planning and Economic Mathematics. He is appointed as a visiting scholar of Pennsylvania State System of Higher Education (Slippery Rock Campus) during September 2011 – February 2012. He has served as a reviewer for several international journals, such as Computers and Mathematics with Applications, Grey Systems: Theory and Applications, Journal of Management Science & Statistical Decision.

Dr. Yao is responsible of National Natural Science Foundation of China, Humanities and Social Sciences Foundation of Ministry of Education of China and Philosophical, Social Science Foundation of Higher Education of Jiangsu Province of China and Sciences Foundation of Nanjing University of Information Science and Technology. He also participated over 20 other foundations.

He has published over 40 research papers. His research interests are mainly in the areas of grey systems theories, fuzzy decision making and energy economics. He was awarded a Distinguished Doctoral Dissertation of Nanjing University of Aeronautics and Astronautics in 2010.

Contents

Chapter 1
Introduction

In a real life decision making, the experience and judgment of experts (or decision makers) represent an effective way to resolve non-structural, or qualitative, or combined qualitative and quantitative decision problems (Saaty, 1980; Xu, 1988). Pairwise comparisons, also known as preference relations, are often used by decision makers to compare a set of decision alternatives with respect to a pre-determined criterion. An ideal process of decision making might wish that each individual reason rationally and the group of all the decision makers cooperates harmonically. In other words, the ideal scenario is that the judgment of each individual decision maker is consistent, and the group of the decision makers reaches good consensus. Then, on the basis of the ideal consistency and consensus, the total amount of information of the individual judgments is aggregated so that the optimized decision can be obtained. However, in reality there is so much complex, uncertain information involved, while the available knowledge is limited and human rationality is bounded. Therefore, it is hard for experts to present precise, consistent judgments; and it is difficult for them to achieve an optimized decision making. Consequently, experts have to rely on incomplete, uncertain, approximately consistent judgments to reach their locally optimal decisions instead of possibly global optimal alternatives. Because of this reason, the theory of expert decision making under uncertainty represents an important scientific focus in the area of decision analysis.

1.1 The Connotation of Uncertainty

In complex and large scale decision makings, due to the complexity involved with the problem itself and the objective environment, and due to the uncertainty of subjective judgments of the decision makers, there are differences among the decision makers in terms of their background knowledge, experience, ability, and preference. The individual and collective judgment information (or knowledge) of the decision makers have the following special features:

(1) The judgment information themselves contain uncertainty, such as fuzziness, randomness, grayness, etc.
(2) The judgment information contains impreciseness. For example, the uncertainty contained in the judgment information itself necessarily can lead

Z. Gong et al.: Uncertain Fuzzy Preference Relations, STUDFUZZ 281, pp. 1–4.
springerlink.com © Springer-Verlag Berlin Heidelberg 2013

to imprecise of the judgment information. In a sense, the imprecise information stands for objective understandings and analyses of the complex problem of decision making.

(3) Each individual judgment contains limitation. That is, there always are limited rationality, prejudice or bias, and preference in each individual judgment; and

(4) There are complexity and conflict in the collective judgment. That is, there are always differences in terms of knowledge, interests, and abilities among all decision makers. Consequently, narrowing differences, expanding the common ground, and reaching the full and unanimous agreement regarding all the possible options and eventually making the right decisions, are among the research issues of decision making analysis.

1.2 The Research Scope of Uncertain Judgment Knowledge

The uncertainty existing in judgment knowledge is characterized by incompleteness and "intuitiveness". The incompleteness of judgment knowledge is reflected in either the incomplete knowledge of the parameters or the incomplete data of decision making, while the "intuitiveness" of judgment knowledge includes fuzzy intervals, which are denoted by continuous intervals within the closed interval [0,1] (Zadeh, 1965; 1975), triangular fuzzy numbers, which are denoted by triads (the degrees of the minimum membership, medium membership, and maximum membership), trapezoidal fuzzy numbers (the degree intervals of the minimum membership and the maximum membership), linguistic fuzzy numbers, such as outstanding, good, poor, etc., and intuitionistic fuzzy numbers (Atanassov, membership degree, non-membership degree, and degree of hesitation). The "intuitiveness" of judgment knowledge is usually used to make intuitive judgments and rational choices in decision makings.

1.3 A Historical Evolution of Uncertainty Fuzzy Sets Theory

In 1965, Zadeh put forward the theory of fuzzy sets, and started a new era of quantitative analysis of qualitative problems. In 1975, Zadeh advanced his theory of fuzzy sets to include interval fuzzy sets by generalizing an ordinary degree of fuzzy membership to an uncertain interval. In the theory of expert decision making, the uncertainty knowledge is correspondingly extended to interval preferences, such as interval values, triangular values, trapezoidal values, and linguistic values from the situation of crisp numbers. In 1982, a thesis entitled "The Control Problem of Grey System", by Deng, symbolized the birth of grey systems theory (Deng, 1998; Liu and Lin, 2010). In grey decision making, grey preferences are employed to represent the judgment intervals in a fixed closed interval $[a,b] \subseteq (-\infty,+\infty)$ of the experts involved. In 1986, the concept of intuitionistic fuzzy sets is proposed (Atanassov, 1986). This concept is not only a generalization of that of the classic fuzzy sets, but also a generalization of that of interval fuzzy sets. Many scholars regard the preference information of intervals, grey and intuitionistic fuzzy sets as preference information of the intuition (Dubois and Gottwald, 2005). The intuitive preference information is virtually an intuitionistic (direct) judgment of the subjective thinking. In 1989, (Atanassov and Gargov, 1989) constructed a relationship between intuitionistic fuzzy sets and interval fuzzy sets so that the study of uncertain preferences is moving forward along with the development of fuzzy set theories. So, the following historical path of development has been observed:

Crisp number preferences → interval preferences → triangular fuzzy number preferences
→ trapezoidal fuzzy number preferences → linguistic fuzzy number preferences
→ two-tuple linguistic fuzzy number preferences
→ intuitionistic fuzzy number preferences.

1.4 Research Contents of Uncertain Preference Relations

Consistent judgment is the foundation of rational choices. However, a large body of facts proves that inconsistency is more likely to happen in complex and uncertain environments (Arrow, 1951; 1963). In the fields of welfare economics and decision making, the debates about consistency and inconsistency have been ongoing for over 50 years, leading to the theoretical developments in social choice and group decision making theory (Arrow, 1951; 1963; Sen, 1964; 1970; Fishburn, 1973). Studies along these lines indicate that

(1) The research of consistency based on intuitive judgment is the foundation of fusing judgment information and seeking optimum alternatives.
(2) Intuitive judgment simulates the thinking of humans; it denotes that inconsistency exists under specific conditions and within specific ranges.
(3) The information of intuitive judgment suffers from incompleteness and uncertainty. It implies that in a real life situation of decision making, it is hard to obtain the optimization alternative. And
(4) The information of intuitive judgment is aggregated by soft computing techniques, which signifies that one can establish satisfactory and/or approximative solutions corresponding to the practical needs, representing the ideal choice of the intuitive judgment among others.

In this book, from the point of the view of the basic theory of fuzzy sets and its various generalizations, we investigate preference relations of uncertainty, including consistency theories, aggregation theories, modeling methods of selecting satisfactory alternatives, and the corresponding intelligent algorithms. In terms of applications, we discuss the feasibility and rationality of the theoretical studies by employing real-life data and case analyses. The focus of this book is the consistency theory and the relevant optimization models of intuitionistic fuzzy preference relations.

In terms of relevant literature, Baets et al. (Baets, Meyer, Schuymer, and Jenei, 2006; Baets and Meyer, 2005; Díaz, Montes and Baets, 2004; Díaz, Baets, and·Montes, 2010; Freson, Meyer, and Baets, 2010) proposed frameworks for studying the transitivity property of preference models, including reciprocal relations and additive fuzzy preference. They (Baets and Meyer, 2008; Baets, Walle, and Kerre, 1995; Baets, and Fodor, 1997) also provided means for the construction of fuzzy strict preferences, indifference, and incomparability relations based on large fuzzy preference relations in an axiomatic framework. Additionally, Rademaker and Baets (2011) discussed some new approaches to the aggregation of preferences, while keeping the naturally existing transitivity property of strict preferences intact. (Fodor and Roubens, 1994; Fodor, 1992; Fodor and Orlovski, 1998) established axiomatic approaches to the definition of fuzzy strict preferences, indifference, and incomparability associated with fuzzy preference relations. (Bisdorff and Roubens, 2004; Ovchinnikov and Roubens, 1992) considered extensions of some classical rational axioms introduced in the conventional choice theory to the circumstances of valued preference relations.

At this junction, we like to mention that all the preference relations studied in this book are different from those considered by Baets et al. In our models of preference relations, we emphasize on judgment matrices as presented by experts as preference relations, while in the models considered by Baets et al. are binary relations.

Chapter 2
Relevant Theories of Reciprocal Preference and Fuzzy Preference Relations

In multiple-attribute decision making, fuzzy theory is very helpful for dealing with the fuzziness widely existing human judgments. Usually, decision makers obtain their subjective opinions by comparing each pair of decision alternatives and then constructing a rational (consistent) judgment matrix, known as a preference relation (PR) (Saaty, 1980; Xu, 1988) to deal with the ranking of alternatives $X = \{x_1, x_2, \cdots, x_n\}$. The research of preference relations includes two main aspects. One is about the consistency properties and consistency check theories of preference relations. The other is about the priorities of the complementary preference relations. There are abundant results about these two aspects (Xu, 2004b; Jiang and Fan, 2008; Gong, 2006). In this chapter, we only briefly discuss some of the fundamental results of preference relations (Gong and Liu, 2006a; 2006e; Gong and Bai, 2008; Gong, 2008), which form the theoretical basis of the later chapters.

This chapter is organized as follows. In Section 2.1, we introduce the concepts of multiplicative preference relations and complementary preference relations and corresponding scales. Then, we detail the properties of these two kinds of PRs. In Section 2.2, we construct an optimization priority model for incomplete complementary preference relations. Section 2.3 studies methods of ranking fuzzy judgment relations, representing an active research area of the past two decades. A short Section 2.4 concludes this chapter.

2.1 Preference Relations and Their Scales

For simplicity, let $P = \{1, 2, \ldots, p\}$ be the set of all natural numbers from 1 to p. In particular, we have $N = \{1, 2, \ldots, n\}$ and $M = \{1, 2, \ldots, m\}$.

In decision making, pairwise comparisons are often used by the decision makers to compare a set $X = \{x_1, x_2, \cdots, x_n\}$ of decision alternatives with respect to a given criterion. Different decision makers may have different preferences. Usually, there are two main preferences used to quantify comparative judgments. One is the reciprocal preference relation (Judgment Matrix) (Saaty, 1980; Xu, 1988), as

Z. Gong et al.: Uncertain Fuzzy Preference Relations, STUDFUZZ 281, pp. 5–18.
springerlink.com © Springer-Verlag Berlin Heidelberg 2013

initially proposed by Saaty in 1980. For a set $X = \{x_1, x_2, \cdots, x_n\}$ of decision alternatives, the preference information of pairwise comparisons with respect to a single criterion is represented numerically using a positive reciprocal matrix

$$A = \begin{pmatrix} a_{11} & a_{12} & \cdots & a_{1n} \\ a_{21} & a_{22} & \cdots & a_{2n} \\ \vdots & \vdots & \vdots & \vdots \\ a_{n1} & a_{n2} & \cdots & a_{nn} \end{pmatrix} \tag{2.1}$$

on the scale of 1 – 9 (Saaty, 1980). The (i,j) entry a_{ij} estimates the degree or intensity of preference of the ith alternative x_i over the jth alternative x_j, and satisfies $a_{ij}\, a_{ji} = 1$ and $a_{ij} > 0$. Such a matrix is also referred to as a multiplicative preference relation. Particularly, $a_{ij} = 1$ indicates indifference between the ith and the jth alternatives x_i and x_j, while $a_{ij} > 1$ indicates that x_i is preferred to x_j, and $a_{ij} < 1$ that x_j is preferred to x_i. The meanings of scales 1 – 9 are shown in Table 2.1.

Table 2.1 Scales1 – 9

Scales	Meanings
1	x_i is the same as x_j
3	x_i is slightly preferred to x_j
5	x_i is definitely preferred to x_j
7	x_i is strongly preferred to x_j
9	x_i is extremely preferred to x_j
2, 4, 6, 8	The mean values of every two adjacent estimations
Reciprocals	If the degree of preference of alternative x_i over x_j is a_{ij}, then the degree of preference of x_j over x_i is $a_{ji} = a_{ij}^{-1}$.

The concept of complementary preferences, also known as fuzzy preference relations is initially put forward by Orlovsky (Orlovsky, 1978). For a set $X = \{x_1, x_2, \cdots, x_n\}$ of decision alternatives, the preference information of pairwise comparisons with respect to a single criterion is represented numerically using a complementary matrix

$$R=(r_{ij})_{n\times n}=\begin{pmatrix} r_{11} & r_{12} & \cdots & r_{1n} \\ r_{21} & r_{22} & \cdots & r_{2n} \\ \vdots & \vdots & \vdots & \vdots \\ r_{n1} & r_{n2} & \cdots & r_{nn} \end{pmatrix} \tag{2.2}$$

on fuzzy scales of 0.1 – 0.9 (Wang and Xu, 1990), where the entry r_{ij} estimates the degree or intensity of preference of the ith alternative x_i over the jth alternative x_j, and satisfies $r_{ij}+r_{ji}=1$, $r_{ij}>0$. Such a matrix is referred to as a fuzzy preference relation (FPR). Particularly, $r_{ii}=0.5$ indicates indifference between x_i and x_j, while $r_{ij}>0.5$ indicates that x_i is preferred to x_j, and $r_{ij}<0.5$ that x_j is preferred to x_i. The meanings of the scales of 0.1 – 0.9 are shown in Table 2.2.

Table 2.2 Scales between 0.1 – 0.9

0.1-0.9 Scales	Meanings
0.1	x_j is extremely preferred to x_i
0.2	x_j is strongly preferred to x_i
0.3	x_j is definitely preferred to x_i
0.4	x_j is slightly preferred to x_i
0.5	x_i is the same as x_j
0.6	x_i is slightly preferred to x_j
0.7	x_i is definitely preferred to x_j
0.8	x_i is strongly preferred to x_j
0.9	x_i is extremely preferred to x_j
complementary number	If the preference degree or intensity of alternative x_i over x_j is r_{ij}, then the preference degree or intensity of alternative x_j over x_i is $r_{ji}=1-r_{ij}$.

The following details two kinds of preference relations:

Definition 2.1. (Saaty, 1980; Wang and Xu,1990;Wang, 1995; Fan and Jiang, 2001). The matrix $\boldsymbol{A}=(a_{ij})_{n\times n}$, as defined in eq. (2.1), is referred to as a multiplicative preference relation (MPR) if $a_{ii}=1$, $a_{ij}=\frac{1}{a_{ji}}$, $a_{ij}>0$, $i,j\in N$, where $\frac{1}{9}\le a_{ij}\le 9$.

Definition 2.2. (Tanino, 1984). The matrix $\boldsymbol{R} = (r_{ij})_{n \times n}$, as defined in eq. (2.2), is referred to as a fuzzy preference relation (FPR), provided that $r_{ii} = 0.5, r_{ij} + r_{ji} = 1, r_{ij} \geq 0, i, j \in N, i \neq j$, where $0.1 \leq a_{ij} \leq 0.9$.

2.2 Properties of Preference Relations

The properties of preference relations not only show whether or not a preference relation is well constructed, but also test whether or not the thinking of a decision maker is consistent and logical. To this end, let us first discuss the properties of multiplicative preference relations (MPRs).

2.2.1 Properties of MPRs

Definition 2.3 (Saaty, 1980; Wang and Xu,1990). If a multiplicative preference relation $\mathrm{A} = (a_{ij})_{n \times n}$ satisfies $a_{ij} = a_{ik} a_{kj}, i, k, j \in N$, then A is referred to as a completely consistent MPR.

Definition 2.4 (Saaty, 1980; Wang, 1995). Let $\boldsymbol{A} = (a_{ij})_{n \times n}$ be a fuzzy preference relation (FPR), for $i, j, k \in N, i \neq j \neq k$.

(1) For $1 \leq \lambda \leq 9$, if $a_{ij} \geq \lambda$, $a_{jk} \geq \lambda$, then $a_{ik} \geq \lambda$, and
(2) For $1/9 \leq \lambda \leq 1$, if $a_{ij} \leq \lambda$, $a_{jk} \leq \lambda$, then $a_{ik} \leq \lambda$,

then $\boldsymbol{A}$ is said to have the property of general transitivity.

Definition 2.5 (Jiang and Fan, 2008; Wang, 1995). Let $\boldsymbol{A} = (a_{ij})_{n \times n}$ be a multiplicative preference relation (MPR), for $i, j, k \in N, i \neq j \neq k$.

(1) If $a_{ik} \geq 1$ and $a_{kj} \geq 1$ imply $a_{ij} \geq 1$, and
(2) If $a_{ik} \leq 1$ and $a_{kj} \leq 1$ imply $a_{ij} \leq 1$,

then $\boldsymbol{A}$ is said to possess the property of satisfactory transitivity.

Property 2.1 (Jiang and Fan, 2008). Each completely consistent MPR $\boldsymbol{A} = (a_{ij})_{n \times n}$ possesses the properties of general transitivity and satisfactory transitivity.

Let us consider a priority chain $x_{u1} \succ x_{u2} \succ \cdots \succ x_{un}$ in $X = \{x_1, x_2, \cdots, x_n\}$ of n decision alternatives, where x_{ui} denotes the ith alternative in the priority chain, and $x_{ui} \succ x_{uj}$ represents that x_{ui} is preferred or equivalently superior to x_{uj}. For all $i, j, k \in N$, if a priority chain $x_{u1} \succ x_{u2} \succ \cdots \succ x_{un}$ exists in $X = \{x_1, x_2, \cdots, x_n\}$ satisfying $x_{ui} \succ x_{uj}$, and $x_{uj} \succ x_{uk}$, then there must be $x_{ui} \succ x_{uk}$. In other words, for all $i, j, k \in N$, if the elements in a multiplicative preference relation (MPR) satisfy

$a_{ik} \geq (>)1$ and $a_{jk} \geq (>)1$, then there must be $a_{ik} \geq (>)1$. Therefore, the weak transitivity of the MPR is associated with a priority chain in the following sense.

Theorem 2.1. An multiplicative preference relation (MPR) possesses the property of satisfactory transitivity, if and only if there exists a priority chain $x_{u1} \succ x_{u2} \succ \cdots \succ x_{un}$ in the set $X = \{x_1, x_2, \cdots, x_n\}$ of decision alternatives, where x_{ui} denotes the ith alternative in the priority chain, and $x_{ui} \succ x_{uj}$ represents x_{ui} is preferred or equivalent to x_{uj}.

According to Theorem 2.1, if there exists a circulation $x_{ui_0} \succ \cdots \succ \cdots \succ x_{ui_0}$ in the decision alternatives, then the corresponding preference relation on the set $X = \{x_1, x_2, \cdots, x_n\}$ of decision alternatives does not possess the property of transitivity, and the multiplicative preference relation is inconsistent.

2.2.2 Properties of Fuzzy Preference Relations (FPRs)

In the following, we establish such properties (Herrera-Viedma et al., 2004; Xu, 2002) of fuzzy preference relations (FPRs) as additive consistency, multiplicative consistency, general transitivity, satisfactory transitivity, and restricted max-min transitivity. These properties can further reflect the judgment thinking of the decision makers.

First of all, there are two kinds of the completely consistent fuzzy preference relations (FPRs).

Definition 2.6 (Song and Yang, 2003; Xu, 2002). A fuzzy preference relation (FPR) $\boldsymbol{R} = (r_{ij})_{n\times n}$ is said to have additive consistency, if it satisfies $r_{ij} = r_{ik} - r_{jk} + 0.5, i, j, k \in N, \quad i \neq j \neq k$. In his case, this FPR is referred to as an additively consistent FPR.

Definition 2.7 (Song and Yang, 2003). A fuzzy preference relation (FPR) $\boldsymbol{R} = (r_{ij})_{n\times n}$ is said to have multiplicative consistency, if it satisfies $r_{ik} r_{kj} r_{ji} = r_{ij} r_{jk} r_{ki}, i, j, k \in N, i \neq j \neq k$. In this case, this FPR is referred to as a multiplicatively consistent FPR.

As a matter of fact, these two kinds of completely consistent FPRs are derived from the concept of completely consistent MPRs. In particular, completely consistent FPRs and completely consistent MPRs can be transformed into each other by using the following transformations.

Lemma 2.1 (Song and Yang, 2003). Let $\boldsymbol{A} = (a_{ij})_{n\times n}$ and $\boldsymbol{R} = (r_{ij})_{n\times n}$ be a multiplicative and a fuzzy preference relation, respectively. Then the following hold true:

$$r_{ij} = \frac{1}{2}(1 + \tfrac{4}{5}\log_9^{a_{ij}}); \ a_{ij} = 3^{5(r_{ij}-0.5)}, i, j \in N \tag{2.3}$$

$$r_{ij} = (1 + a_{ji})^{-1}; a_{ij} = r_{ij} r_{ji}^{-1}, i, j \in N \tag{2.4}$$

Theorem 2.2 (Song and Yang, 2003)**.** Completely consistent MPRs and the two kinds of completely consistent FPRs can be transformed into each other by using eqs. (2.3) and (2.4).

Definition 2.8 (Jiang and Fan, 2008)**.** Let $\boldsymbol{R} = (r_{ij})_{n\times n}$ be a fuzzy preference relation (FPR). For any $i, j, k \in N, i \neq j \neq k$,

(1) If when $0.5 \le \lambda < 1$, $r_{ij} \ge \lambda, r_{jk} \ge \lambda$ imply $r_{ik} \ge \lambda$, and
(2) If when $0 < \lambda \le 0.5$, $r_{ij} \le \lambda, r_{jk} \le \lambda$ imply $r_{ik} \le \lambda$,

then the FPR $\boldsymbol{R}$ is said to have general transitivity.

Definition 2.9 (Jiang and Fan, 2008)**.** Let $\boldsymbol{R} = (r_{ij})_{n\times n}$ be a fuzzy preference relation (FPR), For any $i, j, k \in N, i \neq j \neq k$,

(1) If $r_{ij} \ge 0.5$ and $r_{jk} \ge 0.5$ imply $r_{ik} \ge 0.5$, and
(2) If $r_{ij} \le 0.5$ and $r_{jk} \le 0.5$ imply $r_{ik} \le 0.5$,

then $\boldsymbol{R}$ is said to have satisfactory transitivity.

A set $X = \{x_1, x_2, \cdots, x_n\}$ of decision alternatives has a priority chain, if the decision alternatives can be ordered as follows $x_{u1} \succ x_{u2} \succ \cdots \succ x_{un}$, where x_{ui} denotes the *i*th alternative in the priority chain, and $x_{ui} \succ x_{uj}$ represents x_{ui} is preferred or equivalently superior to x_{uj}. For any $i, j, k \in N$, if a priority chain in $X = \{x_1, x_2, \cdots, x_n\}$ exists such that $x_{ui} \succ x_{uj}$, and $x_{uj} \succ x_{uk}$, then there must be $x_{ui} \succ x_{uk}$. In other word, for any $i, j, k \in N$, if the elements in the fuzzy preference relation (FPR) satisfy $a_{ik} \ge (>)1$ and $a_{jk} \ge (>)1$, then $a_{ik} \ge (>)1$ must holds true. Therefore, the weak transitivity of a FPR is associated with a priority chain according to the following theorem.

Theorem 2.3. A fuzzy preference relation (FPR) has satisfactory transitivity, if and only if there exists a priority chain $x_{u1} \ge x_{u2} \ge \cdots \ge x_{un}$ in the set $X = \{x_1, x_2, \cdots, x_n\}$ of decision alternatives, where x_{ui} denotes the *i*th alternative in the priority chain, and $x_{ui} \ge x_{uj}$ that x_{ui} is preferred or equivalently superior to x_{uj}.

According to Theorem 2.3, if there is a circulation $x_{ui_0} \ge \cdots \ge \cdots \ge x_{ui_0}$ in the decision alternatives, then the corresponding preference relation on the set $X = \{x_1, x_2, \cdots, x_n\}$ of decision alternatives does no have the transitivity property, and the multiplicative preference relation (MPR) is inconsistent.

Definition 2.10 (Tanino, 1984; Herrera-Viedma and Herrera et al., 2004). A fuzzy preference relation (FPR) $\boldsymbol{R} = (r_{ij})_{n\times n}$ is said to have restricted max-max transitivity, if $r_{ij} \geq 0.5$, $r_{jk} \geq 0.5$ imply $r_{ik} \geq max\{r_{ij}, r_{jk}\}$, where $i, j, k \in N, i \neq j \neq k$.

Definition 2.11 (Tanino, 1984; Herrera-Viedma and Herrera et al., 2004). A fuzzy preference relation (FPR) $\boldsymbol{R} = (r_{ij})_{n\times n}$ is said to have restricted max-min transitivity, if $r_{ij} \geq 0.5$, $r_{jk} \geq 0.5$ imply $r_{ik} \geq min\{r_{ij}, r_{jk}\}$, where $i, j, k \in N, i \neq j \neq k$.

Property 2.2 (Tanino, 1984; Herrera-Viedma and Herrera et al., 2004). Each additively consistent FPR has general transitivity, satisfactory transitivity, restricted max-max transitivity, and restricted max-min transitivity.

In real-life decision making situations, it is hard for the decision makers to provide their preference relations to satisfy the afore-mentioned properties. Consequently, the available preference relations have to be adjusted many rounds in order to be useful. Adjusting inconsistent preference relations is one of the hot topics in the research of preference relations. For more details, please refer to literatures, see, for instance, (Jiang and Fan, 2008; Fan and Jiang, 2001;Tanino, 1984; Herrera-Viedma et al., 2004; Xu, 2002; Song and Yang, 2003; Xu, 2008).

2.3 Priority Method of Fuzzy Preference Relations with Incomplete Information

In some practical situations, due to either the uncertainty of objective things, the vague nature of human judgment, or high orders of the preference relations, the decision makers may obtain a preference relation with its entries being incomplete (Wang and Xu, 1990;Xu, 2004c; 2004d;2005a; 2006; Wang, 1995; Takeda, 2001; Nishizawa, 1997; Carmone et al., 1997; Kim and Han, 1999; Hu and Tsai, 2006; Kim and Ahn, 1997; 1999; Kim et al., 1998; 1999; Ahn et al., 2000). If some of the entries of a preference relation cannot be provided by a decision maker, then the preference relation is said to be incomplete. Only when there are either direct or indirect comparisons of every two decision alternatives, can one utilize the finite information to rank all the alternatives (Xu, 1988).

For any $i, j \in N$, let s_{ij} be the (i,j) entry of the preference relation $S = (s_{ij})_{n\times n}$. Define

$$\delta_{ij} = \begin{cases} 1, & s_{ij} \neq 0 \\ 0, & s_{ij} = 0 \end{cases}$$

where $s_{ij} = 0$ denotes that the entry s_{ij} cannot be determined. The matrix $\Delta = (\delta_{ij})_{n\times n}$ is referred to as the indicator matrix of S . According to graph theory (Li, 1988;Yang, 1989), the implied relationship in the matrix Δ can be denoted by means of a directed graph G with n vertices $v_1, v_2, \cdots, v_n$. In particular, $\delta_{ij} = 1$

if and only if there exists a directed arc $\overrightarrow{v_i v_j}$ in the directed graph G going from v_i to v_j. We say that a directed graph G is strongly connected if, for each pair of vertices v_j and v_k with $j \neq k, j,k \in N$, there is a directed path $\overrightarrow{v_j v_{k_1}}, \overrightarrow{v_{k_1} v_{k_2}}, \cdots, \overrightarrow{v_{k_{r-1}} v_k}$ connecting v_j to v_k. Let $\Delta = (\delta_{ij})_{n\times n}$ be the indicator matrix of $S = (s_{ij})_{n\times n}$, and G the corresponding graph. If $s_{ij} \neq 0$, obviously, $s_{ji} \neq 0$, then $\delta_{ij} = \delta_{ji} = 1$, which implies that there is a piece of comparison information between the decision alternatives X_i and X_j. That is, there exists an arc in G going from v_i to v_j, and an arc going from v_j to v_i; If $s_{ij} = 0$, we have $\delta_{ij} = 0$, which implies that there is no comparison information between the alternatives X_i and X_j. For each pair of decision alternatives X_i and X_j with $i \neq j$, when there exists either direct or indirect comparison information between X_i and X_j, there must be a path from v_i to v_j. That is, the graph G is strongly connected. On the contrary, if the graph G is not strongly connected, which implies that there is no path from v_{i_0} to v_{j_0} for some $i_0, j_0 \in N$. In other way, it can be said that there is no direct or indirect comparison information between X_{i_0} and X_{j_0} so that X_{i_0} and X_{j_0} cannot be compared due to loss of information. If $n \geq 2$, a matrix $B_{n\times n}$ is said to be reducible if there is an $n \times n$ permutation matrix P such that $PBP^T = \begin{pmatrix} A_{11} & 0 \\ A_{21} & A_{22} \end{pmatrix}$, where A_{11} is of order l, $1 \leq l \leq n-1$. If no such P exists then B is said to be irreducible. An irreducible matrix and its directed graph are related in the following fashion (Li, 1988; Yang, 1989):

A square matrix is irreducible if and only if its directed graph is strongly connected.

In consequence, we obtain the following conclusion:

Let R be an incomplete preference relation, Δ the indicator matrix of R, and G the directed graph of Δ. Then G is strongly connected $\Leftrightarrow$ Δ is irreducible $\Rightarrow$ all the decision alternatives can be ranked by R utilizing some approaches (Li, 1988; Yang, 1989).

2.3.1 A Least Square Model for Collective Preference Relations with Incomplete Information

This subsection is based on (Gong, 2008).

Let $X = \{X_1, X_2, \cdots, X_n\}$ be a set of decision alternatives, and $d = \{d_1, d_2, \cdots, d_m\}$ a set of decision makers. For simplicity, let us denote

$N = \{1, 2, \cdots, n\}$, $M = \{1, 2, \cdots, m\}$. The preferences of the decision makers on X are described by fuzzy preference relations as follows:

$$R_s = \begin{pmatrix} r_{11s} & r_{12s} & \cdots & r_{1ns} \\ r_{21s} & r_{22s} & \cdots & r_{2ns} \\ \cdots & \cdots & \cdots & \cdots \\ r_{n1s} & r_{n2s} & \cdots & r_{nns} \end{pmatrix},$$

where r_{ijs} denotes the degree of preference of decision alternative X_i over X_j as presented by the sth decision maker, for each $s \in M$.

For reasons either within or beyond control, the decision makers may present incomplete information. Therefore, for $i_0, j_0 \in N, s_0 \in M$, if $r_{i_0 j_0 s_0}$ is incomplete, and we write $r_{i_0 j_0 s_0} = -$. Suppose that there are d_{ij} decision makers who present their degrees of preference of the decision alternative X_i over X_j, which also implies that there are d_{ji} decision makers who present their degrees of preference of decision alternative X_j over X_i because of $r_{ijs} = 1 - r_{jis}$. Therefore, we have $d_{ij} = d_{ji}$, it is clear that d_{ij} satisfies $0 \le d_{ij} \le m$.

Let us look at an ideal case. For a decision-making problem, all the individual judgments of the decision makers are unanimous. That is, all the decision makers provide the same fuzzy preference relations. As we have seen, if $V = (v_1, v_2, \cdots, v_n)^T$ is the priority vector of multiplicative consistent fuzzy preference relation $R_s = (r_{ijs})_{n \times n}$, then

$$r_{ijs} = v_i / (v_i + v_j), i, j \in N, s \in M.$$

Moreover, for $r_{ijs} + r_{jis} = 1$, we have

$$r_{ijs} v_j = r_{jis} v_i \tag{2.5}$$

So, for any $i, j \in N, s \in M$, the following equation holds true:

$$r_{ijs} v_j - r_{jis} v_i = 0 \tag{2.6}$$

However, in the general case, the decision makers may disagree with each other so that the previous equation does not hold true in general. Let $\varepsilon_{ijs} = (r_{ijs} v_j - r_{jis} v_i)^2$. For any $i, j \in N, s \in M$, the smaller ε_{ijs} is, the higher the consensus of the judgments of the decision makers. Our objective here is to produce an optimal priority vector $V = \{v_1, v_2, \cdots, v_n\}^T$ by minimizing the error ε_{ijs} for all $i, j \in N, s \in M$. Therefore, to this end, we construct the following constrained optimization model, where it is supposed that all the decision makers have the same weights:

$$\begin{cases} \min\ g(v) = \sum_{i=1}^{n}\sum_{j=1}^{n}\sum_{s=1}^{m}(r_{jis}v_i - r_{ijs}v_j)^2 \\ s.t. \sum_{i=1}^{n} v_i = 1, v_i > 0, i \in N. \end{cases} \tag{2.7}$$

However, when some of the decision makers present incomplete information, that is, there are some $i_0, j_0 \in N, s_0 \in M$ such that $r_{i_0 j_0 s_0} = -$. By deleting all the incomplete information, we obtain the following optimization model:

$$\begin{cases} \min\ g(v) = \sum_{i=1}^{n}\sum_{j=1}^{n}\sum_{l=1}^{d_{ij}}(r_{jil}v_i - r_{ijl}v_j)^2 \\ s.t. \sum_{i=1}^{n} v_i = 1, v_i > 0, i \in N. \end{cases} \tag{2.8}$$

where $r_{ijl} \neq -$ denotes that the *lth* decision maker presents his degree of preference of decision alternative X_i over X_j, $i, j \in N, l \in \{1,\ldots,d_{ij}\}$.

Consider a *Lagrange* function

$$L(v,\lambda) = \sum_{i=1}^{n}\sum_{j=1}^{n}\sum_{l=1}^{d_{ij}}(r_{jil}v_i - r_{ijl}v_j)^2 + 4\lambda(\sum_{i=1}^{n} v_i - 1) \tag{2.9}$$

By letting $\frac{\partial L}{\partial v_j} = 0$, we have

$$\sum_{i=1}^{n}(\sum_{l=1}^{d_{ji}} r_{ijl}^2 v_j - \sum_{l=1}^{d_{ji}} r_{jil} r_{ijl} v_i) + \lambda = 0, j \in N \tag{2.10}$$

Its equivalent matrix form is given as follows:

$$QV + \lambda e = 0 \tag{2.11}$$

where

$$Q = (q_{ij})_{n\times n} = \begin{pmatrix} \sum_{i=1,i\neq 1}^{n}\sum_{l=1}^{d_{1i}} r_{i1l}^2 & -\sum_{l=1}^{d_{12}} r_{12l}r_{21l} & \cdots & -\sum_{l=1}^{d_{1n}} r_{1nl}r_{n1l} \\ -\sum_{l=1}^{d_{21}} r_{12l}r_{21l} & \sum_{i=1,i\neq 2}^{n}\sum_{l=1}^{d_{2i}} r_{i2l}^2 & \cdots & -\sum_{l=1}^{d_{2n}} r_{2nl}r_{n2l} \\ \cdots & \cdots & \cdots & \cdots \\ -\sum_{l=1}^{d_{n1}} r_{1nl}r_{n1l} & -\sum_{l=1}^{d_{n2}} r_{n2l}r_{2nl} & \cdots & \sum_{i=1,i\neq n}^{n}\sum_{l=1}^{d_{ni}} r_{inl}^2 \end{pmatrix},$$

$V = (v_1, v_2, \cdots, v_n)^T, e = (1,1,\cdots,1)^T$.

For any $i \neq j$, $q_{ij} \neq 0$ if and only if $d_{ij} \neq 0$. Because $d_{ij} \neq 0$ denotes that there is at least one decision maker who presents his comparison information between decision alternatives X_i and X_j, $q_{ij} \neq 0$ can also be explained that there is comparison information between the alternatives X_i and X_j. So, it is definite that Q implies that there are $\Longleftrightarrow$ decision makers who present their information comparing alternatives X_i and X_j. So, for any Q, if there exist both direct and indirect comparison information between X_i and X_j, that is, the graph of is strongly connected and the indicator matrix of is irreducible, then we can rank all the alternatives in the set $X = \{X_1, X_2, ..., X_n\}$.

Therefore, we have the following conclusion:

Lemma 2.2. The graph of Q is strongly connected $\Leftrightarrow$ the indicator matrix of Q is irreducible $\Rightarrow$ all the decision alternatives can be ranked by using the model in eq. (2.8).

Lemma 2.3. If the indicator matrix of Q is irreducible, then Q is a positive define matrix, and Q is invertible.

Proof. Because

$$
\begin{aligned}
g(v) &= \sum_{i=1}^{n}\sum_{j=1}^{n}\sum_{l=1}^{d_{ij}} (a_{jil}v_i - a_{ijl}v_j)^2 \\
&= \sum_{i=1}^{n}\sum_{j=1}^{n}\sum_{l=1}^{d_{ij}} (a_{jil}v_i^2 - 2a_{jil}a_{ijl}v_iv_j + a_{ijl}v_j^2) \\
&= \sum_{i=1}^{n}\sum_{j=1}^{n}\sum_{l=1}^{d_{ij}} a_{jil}^2v_i^2 + \sum_{i=1}^{n}\sum_{j=1}^{n}\sum_{l=1}^{d_{ij}} a_{ijl}^2v_j^2 - \sum_{i=1}^{n}\sum_{j=1}^{n}\sum_{l=1}^{d_{ij}} 2a_{jil}a_{ijl}v_iv_j \\
&= 2\sum_{i=1}^{n}\sum_{j=1}^{n}\sum_{l=1}^{d_{ij}} a_{ijl}^2v_j^2 - 2\sum_{i=1}^{n}\sum_{j=1}^{n}\sum_{l=1}^{d_{ij}} a_{jil}a_{ijl}v_iv_j \\
&= 2\sum_{j=1}^{n}\Big(\sum_{i=1, i\neq j}^{n}\sum_{l=1}^{d_{ij}} a_{ijl}^2\Big)v_j^2 - 2\sum_{j=1}^{n}\sum_{i=1, i\neq j}^{n}\sum_{l=1}^{d_{ij}} a_{jil}a_{ijl}v_iv_j \\
&= 2V^TQV
\end{aligned}
$$

and because the indicator matrix of Q is irreducible, according to Lemma 2.2, we can utilize $A_s, s \in M$ and the model in eq. (2.8) to rank all the decision alternatives. That is to say, the priority vector V exists. For $V > 0$, there must be $g(v) > 0$, because $Q^T = Q$. So Q is a positive define matrix. Therefore, Q^{-1} exists. QED.

From eq. (2.11), we have

$$V = -\lambda Q^{-1} e \tag{2.12}$$

and

$$e^T V = -\lambda e^T Q^{-1} e.$$

Since $e^T V = 1$, we have

$$1 = -\lambda e^T Q^{-1} e.$$

That is

$$\lambda = -1 / e^T Q^{-1} e.$$

By eq. (2.12), we have

$$V = Q^{-1} e / e^T Q^{-1} e.$$

Thus we establish the following theorem:

Theorem 2.4. Let $A_s = \left(a_{ijs}\right)$, for $i, j \in N, s \in M$, be a profile of incomplete fuzzy preference relations such that its indicator matrix Q is irreducible. Then the optimal solution of the model in eq. (2.8) can be written as follows:

$$V = Q^{-1} e / e^T Q^{-1} e \tag{2.13}$$

In fact, $Q = (q_{ij})_{n \times n}$ can be regarded as a measurement of the degree of incompleteness of preference relations. In other words, for any $i, j \in N, i \neq j$, $d_{ij} = 0$ denotes that no one presents the degree of preference for the decision alternative X_i over X_j. So, $q_{ij} = 0$; and $d_{ij} = m, m \geq 1$, indicates that all the decision makers present their degrees of preference of decision alternative X_i over X_j. In this case, $q_{ij} \neq 0$.

We can extend the priority model of incomplete fuzzy preference relations, as established above, to collective group decisions and individual decisions with complete information.

For all $i, j \in N$, if we let $d_{ij} = m$, where d_{ij} is the number of the decision makers who present their degrees of preference of decision alternative X_i over X_j, then we can establish a collective priority model of the fuzzy preference relations as presented by multiple decision makers with complete information as follows (Xu, 2002):

$$\begin{cases} \min\ g(v) = \sum_{i=1}^{n}\sum_{j=1}^{n}\sum_{s=1}^{m}(r_{jis}v_i - r_{ijs}v_j)^2 \\ s.t. \sum_{i=1}^{n} v_i = 1, v_i > 0, i \in N. \end{cases} \tag{2.14}$$

The priority vector of this model is given in eq. (2.13).

For any $i, j \in M$, if we let $d_{ij} = 1$, we can also obtain a priority model of an individual fuzzy preference relation with complete information (Xu, 2002):

$$\begin{cases} \min\ g(v) = \sum_{i=1}^{n}\sum_{j=1}^{n}(r_{ji}v_i - r_{ij}v_j)^2 \\ s.t. \sum_{i=1}^{n} v_i = 1, v_i > 0, i \in N. \end{cases} \tag{2.15}$$

The priority vector of this model is also given by eq. (2.13).

2.3.2 Numerical Example

Suppose that three decision makers provide the following incomplete preference relations $\{R_1, R_2, R_3\}$ on a set of four decision alternatives $X = \{X_1, X_2, X_3, X_4\}$, and that all the decision makers have the same weights, where

$$R_1 = \begin{pmatrix} 0.5 & 0.2 & - & - \\ 0.8 & 0.5 & 0.4 & - \\ - & 0.6 & 0.5 & 0.7 \\ - & - & 0.3 & 0.5 \end{pmatrix}; R_2 = \begin{pmatrix} 0.5 & - & 0.7 & - \\ - & 0.5 & - & 0.9 \\ 0.3 & - & 0.5 & - \\ - & 0.1 & - & 0.5 \end{pmatrix}; R_3 = \begin{pmatrix} 0.5 & 0.4 & - & - \\ 0.6 & 0.5 & 0.3 & - \\ - & 0.7 & 0.5 & - \\ - & - & - & 0.5 \end{pmatrix}.$$

Step1: By applying the model in eq. (2.8), we get the following positive define matrix Q:

$$Q = \begin{pmatrix} 1.09 & -0.4 & -0.21 & 0 \\ -0.4 & 1.06 & -0.45 & -0.09 \\ -0.21 & -0.45 & 0.83 & -0.21 \\ 0 & -0.09 & -0.21 & 1.3 \end{pmatrix}.$$

Step2: According to eq. (2.13), the priority vector of group decision is

$$(v_1, v_2, v_3, v_4)^T = (0.2360, 0.3036, 0.3349, 0.1255)^T.$$

Thus the ranking of group decisions is $X_3 \succ X_2 \succ X_1 \succ X_4$.

2.4 Conclusions

In this chapter, after introducing two kinds of preference relations: the MPR and the FPR, and their corresponding scales, we detailed the properties of these preference relations. In Section 2.3, by using the transformation relation between multiplicative preference relations and fuzzy preference relations, we developed a least-square priority model of collective preference relations with incomplete information, while developing the condition for the solution to exist. Then, this model was generalized to the cases of collective preference relations and individual preference relations of complete information. At the end, we constructed a numerical example to demonstrate the fundamental idea presented in this chapter. Because the priority model proposed here is based on multiplicative consistency of fuzzy preference relations, we can also develop a least logarithmic square approach to fuzzy preference relations based on additive consistency by using the same method.

Chapter 3
Complementary Preference Relations of Interval Fuzzy Numbers

In any environment that is complex and involves uncertainty, due to either the uncertainty involved with the objective things or the vague nature of human judgments or both, decision makers might prefer imprecise information of judgment that takes the form of uncertainty numbers such as interval fuzzy numbers. Because it is hard to produce any accurate estimate, interval estimations are more preferable than crisp numbers. The lower bound of an interval estimate stands for the minimum judgment, conservative estimation, and/or pessimistic outlook, and the upper bound the maximum judgment and/or optimistic outlook. The analysis of interval preference originates from Zadeh's (1975) interval-valued fuzzy sets (IVFS).

Usually, there are two kinds of preference relations involving intervals. One is the so-called reciprocal preference relation of interval fuzzy numbers or IFNRPR for short; and the other complementary preference relation of interval fuzzy numbers or IFNCPR for short (Li and Liu, 2004; Xu and Yang, 1998; Wei et al., 1994; Wang et al., 1995; Zhang, 2003; Wang and Elhag, 2007; Wang and Chin, 2008; Wang, 2006; Wang and Fan, 2007; Wang, Luo et al., 2008; Wang and Elhag, 2006; Wang, Elhag et al., 2006; Xu and Da, 2003). Among the research topics, issues related to consistency and priorities have been the hot topic of discussion of IFNRPRs. Basing on what has been published in the literature, this chapter will investigate the consistency theory and priority method of IFNCPRs. In particular, in Sections 3.2 and 3.3, properties of two different kinds of consistent IFNCPRs will be detailed. In Sections 3.4 and 3.5, two kinds of priority models of the IFNCPR with incomplete information will be constructed and it will be shown that these models are actually the generalization of that of the group IFNCPRs with complete information.

3.1 Basic Concepts

All intervals mentioned in this chapter will be assumed to be positive. For the convenience of our communication, some basic laws of operation of positive interval are given below. For more in-depth discussion, please consult with (Xu and

Z. Gong et al: Uncertain Fuzzy Preference Relations, STUDFUZZ 281, pp. 19–44.
springerlink.com © Springer-Verlag Berlin Heidelberg 2013

Da, 2003a,2003b; Young, 1931; Moore, 1979; Moore and Lodwick, 2003; Gong and Liu, 2006).

Let $M_1 = [l_1, u_1]$ and $M_2 = [l_2, u_2]$ be two intervals. Then the basic arithmetic operations between these intervals, such as addition +, scalar multiplication, multiplication · operations , are defined as follows:

$$[l_1,u_1]+[l_2,u_2]=[l_1+l_2,u_1+u_2] ;$$
$$[l_1,u_1]-[l_2,u_2]=[l_1-u_2,u_1-l_2];$$
$$[l_1,u_1)\cdot[l_2,u_2]= [l_1l_2,u_1u_2];$$
$$\lambda\cdot[l_1,u_1]=[\lambda l_1,\lambda u_1], \lambda>0, \lambda\in R ;$$
$$[l_1,u_1]\div[l_2,u_2]=[l_1/u_2,u_1/l_2];$$
$$lnM_1 = (lnl_1, lnu_1) ; \; exp(M_1)= (expl_1, expu_1) .$$

Each real number $a \in R$ can be denoted as $a=[a,a]$. By $[l_1,u_1]\geq a$, it means that both $l_1 \geq a$ and $u_1 \geq a$.

Definition 3.1 (Wei, Liu et al, 1994; Wang, Liu et al., 1995). A preference relation ***A*** is referred to as interval reciprocal or IFNRPR for short, if the following conditions hold

$$a_{ii}=[1,1] \tag{3.1}$$

$$a_{ijl}a_{jiu}=a_{iju}a_{jil}=1 \tag{3.2}$$

where $a_{ij}=[a_{ijl},a_{iju}]$ denotes the range of the membership degree with which the decision alternative x_i is preferred to another decision alternative x_j, for any $i, j \in N$.

Definition 3.2 (Li and Liu, 2004). A preference relation $\boldsymbol{R}=(r_{ij})_{n\times n}$ is referred to as one of interval fuzzy numbers or IFNCPR for short, if the following conditions hold

$$r_{ii}=[0.5,0.5] \tag{3.3}$$

$$r_{ijl}+r_{jiu}=r_{iju}+r_{jil}=1 \tag{3.4}$$

where $r_{ij}=[r_{ijl},r_{iju}]$ denotes the range of membership degree with which the decision alternative x_i is preferred to the decision alternative x_j, for $i, j \in N$.

3.2 Properties and Priorities of IFNCPRs with Additive Consistency

3.2.1 Properties of IFNCPRs with Additive Consistency

An IFNCPR $\mathbf{R}=(r_{ij})_{n\times n}$ and an IFNRPR $\mathbf{A}=(a_{ij})_{n\times n}$ can be transformed to each other from the point of the view of additive consistency. (This section is based on (Gong and Liu, 2006a, 2006b)).

Theorem 3.1. An IFNCPR $\mathbf{R}=(r_{ij})_{n\times n}$ and an IFNRPR $\mathbf{A}=(a_{ij})_{n\times n}$ can be transformed to each other by the following formulas:

$$r_{ij}=0.5+0.2log_3^{a_{ij}} \tag{3.5}$$

$$a_{ij}=3^{5(r_{ij}-0.5)} \tag{3.6}$$

Proof. Let $\mathbf{A}$ be an IFNRPR. From Eq. (3.5) and $a_{ii}=[1,1]$, we have $r_{ii}=[0.5,0.5]$. And for the reason that

$$a_{ijl}a_{jiu}=a_{iju}a_{jil}=1$$

we have

$$r_{ijl}+r_{jiu}=0.5+0.2log_3^{a_{ijl}}+0.5+0.2log_3^{a_{jiu}}=1$$

Similarly, we can readily obtain that

$$r_{iju}+r_{jil}=1$$

Thus $\mathbf{R}=(r_{ij})_{n\times n}$ is an IFNCPR.

Using the same method, we can transform $\mathbf{R}=(r_{ij})_{n\times n}$ into $\mathbf{A}=(a_{ij})_{n\times n}$ by Eq. (3.6). QED.

Definition 3.3 (Wei, Liu et al, 1994; Wang, Liu et al., 1995). An IFNRPR $\mathbf{A}=(a_{ij})_{n\times n}$ is multiplicatively consistent, if

$$a_{ij}=1/a_{ji} \tag{3.7}$$

$$a_{ij}a_{jk}=a_{jj}a_{ik} \tag{3.8}$$

Eq. (3.8) is equivalent to

$$a_{ijl}a_{jkl}=a_{jjl}a_{ikl} \tag{3.9}$$

$$a_{iju}a_{jku}=a_{jju}a_{iku} \tag{3.10}$$

The definition of additively consistent IFNCPRs can be derived by using Definition 3.3.

From Eq. (3.9), we have

$$0.2log_3^{a_{ijl}} + 0.2log_3^{a_{jkl}} = 0.2log_3^{a_{ikl}}$$

Thus

$$(0.5 + 0.2log_3^{a_{ijl}}) + (0.5 + 0.2log_3^{a_{jkl}}) = 1 + 0.2log_3^{a_{ikl}}$$

That is

$$r_{ijl} + r_{jkl} = r_{ikl} + 0.5 \quad (3.11)$$

Similarly, we can also obtain that

$$r_{iju} + r_{jku} = r_{iku} + 0.5 \quad (3.12)$$

Definition 3.4. An IFNCPR $\mathbf{R} = (r_{ij})_{n\times n}$ is said to have additive consistency, if Eqs. (3.11) and (3.12) hold, for all $i, j, k \in N$.

Definition 3.5. Let $\mathbf{R} = (r_{ij})_{n\times n}$ be an IFNCPR. If for all $i, j, k \in N, i \neq j \neq k$, the following hold:

(1) When $0.5 \leq \lambda < 1$, if $r_{ij} \geq \lambda, r_{jk} \geq \lambda$, then $r_{ik} \geq \lambda$; and
(2) When $0 < \lambda \leq 0.5$, if $r_{ij} \leq \lambda, r_{jk} \leq \lambda$, then $r_{ik} \leq \lambda$

then $\mathbf{R} = (r_{ij})_{n\times n}$ satisfies the general transitivity.

Theorem 3.2. A FNPR with additive consistency has general transitivity.

Proof. Let $\mathbf{R} = (r_{ij})_{n\times n}$ be an additively consistent FNPR. When $0.5 \leq \lambda < 1$, if we let

$$r_{ij} \geq \lambda, r_{jk} \geq \lambda$$

then we have

$$r_{ijl} \geq \lambda, r_{jkl} \geq \lambda, r_{iju} \geq \lambda, r_{jku} \geq \lambda$$

Meanwhile, for the reason that

$$r_{ijl} + r_{jkl} = r_{ikl} + 0.5, r_{iju} + r_{jku} = r_{iku} + 0.5$$

we have

$$r_{ikl} \geq 2\lambda - 0.5 \geq \lambda, r_{iku} \geq 2\lambda - 0.5 \geq \lambda$$

That is

$$r_{ik} \geq \lambda$$

Similarly, when $0 < \lambda \leq 0.5$, if we let $r_{ij} \leq \lambda, r_{jk} \leq \lambda$, then $r_{ik} \leq \lambda$. QED.

Definition 3.6. Let $\mathbf{R} = (r_{ij})_{n \times n}$ be an IFNCPR. If for all $i, j, k \in N, i \neq j \neq k$,

(1) $r_{ij} \geq 0.5, r_{jk} \geq 0.5$ implies $r_{ik} \geq 0.5$; and

(2) $r_{ij} \leq 0.5, r_{jk} \leq 0.5$ implies $r_{ik} \leq 0.5$

then $\mathbf{R} = (r_{ij})_{n \times n}$ is said to have satisfactory consistency.

According to Theorem 3.2 and Definition 3.6, we get the following corollary.

Corollary 3.1. A FNPR with additive consistency also has satisfactory consistency.

An equivalent version of definition 3.6 is the following Definition 3.6'.

Definition 3.6'. If a FNPR has satisfactory consistency, then the corresponding preference relation of the decision alternatives $X = \{x_1, x_2, \cdots, x_n\}$ has transitivity property. That is, there exists a priority chain $x_{u1} \geq x_{u2} \geq \cdots \geq x_{un}$ in $X = \{x_1, x_2, \cdots, x_n\}$, where x_{ui} denotes the ith decision alternative in the priority chain, and $x_{ui} \geq x_{uj}$ represents that x_{ui} is preferred (superior) to x_{uj}. If there is a circulation $x_{ui_0} \geq \cdots \geq \cdots \geq x_{ui_0}$ in a priority chain, then the corresponding preference relation on the set of decision alternatives $X = \{x_1, x_2, \cdots, x_n\}$ is said to have not the transitivity property, and the fuzzy complementary preference relation is said to be inconsistent.

According to this definition, it can be seen that satisfactory consistency is a minimal logical requirement and a fundamental principle of fuzzy preference relations, which reflects a thinking characteristic of man. Therefore, it is very important to set up an approach that can judge whether a given fuzzy complementary preference relation has satisfactory consistency. In the following, we will give a definition of preference matrix.

Definition 3.7. Let $\mathbf{R} = (r_{ij})_{n \times n}$ be a FNPR. $\mathbf{P} = (p_{ij})_{n \times n}$ is the preference matrix of R, if

$$p_{ij} = \begin{cases} 0 & r_{ij} \leq 0.5; \\ 1 & r_{ij} > 0.5 \end{cases}.$$

Theorem 3.3. Let $\mathbf{P} = (p_{ij})_{n \times n}$ be a preference matrix of $\mathbf{R} = (r_{ij})_{n \times n}$, $\mathbf{P}^i$ the ith sub-matrix of $\mathbf{P} = (p_{ij})_{n \times n}$, where $i = 0, 1, \ldots, n$. (That is, deleting one 0 row vector and a corresponding column vector, we get a sub-matrix $\mathbf{P}^1$ of $\mathbf{P}^0 = \mathbf{P}$; …; deleting one 0 row vector and the corresponding column vector of $\mathbf{P}^i$, we get a sub-matrix $\mathbf{P}^{i+1}$ of $\mathbf{P}^i$; …; deleting one 0 row vector and the corresponding column vector of

$\mathbf{P}^{n-1}$, we get a sub-matrix $\mathbf{P}^{n-2}$ of $\mathbf{P}^{n-1}$, and $\mathbf{P}^{n} = (0)$.) For any i = 0, 1, …, n, $\mathbf{R} = (r_{ij})_{n \times n}$ is satisfactorily consistent if and only if there is a 0 row vector in $\mathbf{P}^{i}$.

Proof. Necessity. If $\boldsymbol{R} = (r_{ij})_{n \times n}$ has satisfactory consistency, suppose that we have a set of decision alternatives, $X = \{x_1, x_2, \cdots, x_n\}$ with an associated priority chain $x_{u1} \geq x_{u2} \geq \cdots \geq x_{un}$, where x_{ui} denotes the ith alternative in the priority chain. Since x_{un} is the most inferior decision alternative, we have that $r_{u_{nj}} \leq 0.5, j = 1, \ldots, n$, and so $p_{u_n j} = 0, j = 1, \cdots, n$. That is, the entries of the u_n -th row are all 0. Deleting the u_n -th row and the u_n -th column, we get a sub-matrix $\mathbf{P}^{1}$. At this time, the priority relations of the rest decision alternatives experience no change so that $x_{u_{n-1}}$ is now the most inferior decision alternative of the remaining alternatives. Obviously, in $\mathbf{P}^{1}$, the entries of the row represented by $x_{u_{n-1}}$ are all 0. Deleting the u_n th row and the u_n th column, the u_{n-1} th row and the u_{n-1} th column of **P**, we get $\mathbf{P}^{2}$. Continuing this procedure, we eventually have an (n – 1)th sub-matrix

$$\mathbf{P}^{n-1} = \begin{pmatrix} 0 & 1 \\ 0 & 0 \end{pmatrix} \text{or} \begin{pmatrix} 0 & 0 \\ 1 & 0 \end{pmatrix} \text{or} \begin{pmatrix} 0 & 0 \\ 0 & 0 \end{pmatrix}$$

In $\mathbf{P}^{n-1}$, the 0 row vector is represented by x_{u_2}. Deleting this 0 row vector and the corresponding column, we conclude that $P^n = (0)$ so that the most superior decision alternative x_{u_1} is resulted.

Sufficiency. Let the entries of the u_n th row vector in **P** be 0. So it is obviously that x_{u_n} stands for the most inferior decision alternative. Now deleting the u_n th row and the u_n th column in **P**, we get a sub-matrix $\mathbf{P}^{1}$. Let the 0 row vector be represented by $x_{u_{n-1}}$. Then $x_{u_{n-1}}$ is superior to x_{u_n}. According to this method of comparison, we will eventually obtain the most superior decision alternative x_{u_1}. Thus we have a priority chain $x_{u_1} \geq x_{u_2} \geq \cdots \geq x_{u_n}$, where x_{n_i} denotes the ith superior decision alternative in the set $X = \{x_1, x_2, \cdots, x_n\}$ of all the alternatives. Therefore, **R** has satisfactory consistency. QED.

Actually, according to Theorem 3.3, we can naturally produce a priority algorithm for each satisfactorily consistent FNPR as follows.

Step 1. Construct the preference matrix.
Step 2. Let i = 0.
Step 3. Search the 0 row vector in sub-matrix $\mathbf{P}^{i}$, if the 0 row exists, then the decision alternative that represents this row is denoted $x_{u_{n-i}}$, and go to step 4. Otherwise go to step 5.

Step 4. Delete the 0 row in $\mathbf{P}^i$ (if there are more than one such rows then select a 0 row randomly) and the corresponding column, set $i = i + 1$. If $i = n$, then the decision alternative that represents this row is denoted x_{u_1} (That is, R has satisfactory consistency). End. Otherwise, go to step 3.
Step 5. R is inconsistent. End

Definition 3.8. Let $\mathbf{R} = (r_{ij})_{n\times n}$ be a FNPR. For any $i, j, k \in N, i \neq j \neq k$, R is said to have the property of restricted max-max transitivity, if $r_{ij} \geq 0.5, r_{jk} \geq 0.5$ implies $r_{ikl} \geq max\{r_{ijl}, r_{jkl}\}$, $r_{iku} \geq max\{r_{iju}, r_{jku}\}$.

Definition 3.9. Let $\mathbf{R} = (r_{ij})_{n\times n}$ be a FNPR. For any $i, j, k \in N, i \neq j \neq k$, R is said to have the property of restricted max-min transitivity, if $r_{ij} \geq 0.5, r_{jk} \geq 0.5$ implies $r_{ikl} \geq min\{r_{ijl}, r_{jkl}\}$, $r_{iku} \geq min\{r_{iju}, r_{jku}\}$.

Theorem 3.4. If a FNPR satisfies the property of additive consistency, then it also have the property of restricted max-max transitivity.

Proof. From $r_{ij} \geq 0.5, r_{jk} \geq 0.5$, we get $r_{ijl} \geq 0.5, r_{jkl} \geq 0.5$, $r_{iju} \geq 0.5, r_{jku} \geq 0.5$. So, according to Eqs.(3.11) and (3.12), we have $r_{ijl} + r_{jkl} = r_{ikl} + 0.5$, $r_{iju} + r_{jku} = r_{iku} + 0.5$. Thus $r_{ikl} \geq r_{ijl}, r_{ikl} \geq r_{jkl}$; $r_{iku} \geq r_{iju}, r_{iku} \geq r_{jku}$.

QED.

Corollary 3.2. If a FNPR satisfies the property of additive consistency, then it also satisfies the property of restricted max-min transitivity. QED.

Definition 3.10. Let $\mathbf{R} = (r_{ij})_{n\times n}$ be a FNPR. For any $i, j, p, m, s, n \in N$, R is said to have the property of restricted weak monotonicity, if $r_{ijl} > r_{pml}$, $r_{jnl} > r_{msl}$; $r_{iju} > r_{pmu}$, $r_{jnu} > r_{msu}$ imply $r_{inl} > r_{psl}$; $r_{inu} > r_{psu}$.

Theorem 3.5. If a FNPR satisfies the property of additive consistency, then it also satisfies the property of restricted weak monotonicity.

Proof. Let R $= (r_{ij})_{n\times n}$ be a FNPR. When $r_{ijl} > r_{pml}$, $r_{jnl} > r_{msl}$, we have $r_{ijl} + r_{jnl} > r_{pml} + r_{msl}$. And, for $r_{ijl} + r_{jnl} = r_{inl} + 0.5$, $r_{pml} + r_{msl} = r_{psl} + 0.5$, we have $r_{inl} > r_{psl}$.

When $r_{iju} > r_{pmu}$ and $r_{jnu} > r_{msu}$, the condition $r_{inu} > r_{psu}$ holds similarly. QED.

3.2.2 A Numerical Example

Suppose that a decision maker provides his fuzzy complementary preference relation on a set $X = \{x_1, x_2, x_3, x_4\}$ of decision alternatives as follows:

$$\begin{pmatrix} [0.5,0.5] & [0.2,0.4] & [0.3,0.4] & [0.6,0.9] \\ [0.6,0.8] & [0.5,0.5] & [0.6,0.7] & [0.6,0.7] \\ [0.6,0.7] & [0.3,0.4] & [0.5,0.5] & [0.7,0.7] \\ [0.1,0.4] & [0.3,0.4] & [0.3,0.3] & [0.5,0.5] \end{pmatrix}$$

Step 1: Construct the preference matrix **P** is as follows:

$$\mathbf{P} = \begin{pmatrix} 0 & 0 & 0 & 1 \\ 1 & 0 & 1 & 1 \\ 1 & 0 & 0 & 1 \\ 0 & 0 & 0 & 0 \end{pmatrix}$$

Step 2: Let $\mathbf{P}^0 = \mathbf{P}$.
Step 3: Search the 0 row vector in $\mathbf{P}^0$. Obviously, the entries of the fourth row are all 0. So, x_4 is the most inferior decision alternative.
Step 4: By delete the fourth row and the fourth column in $\mathbf{P}^0$, we get $\mathbf{P}^1$ as follows:

$$\mathbf{P}^1 = \begin{pmatrix} 0 & 0 & 0 \\ 1 & 0 & 1 \\ 1 & 0 & 0 \end{pmatrix}$$

Step 5: Search the 0 row vector in $\mathbf{P}^1$. Obviously, the entries of the first row are all 0. This row is also the second row of **P**. So x_1 is superior to x_4.

Step 6: By deleting the fourth row and the fourth column and the first row and the first column in **P**, we get $\mathbf{P}^2$ as follows:

$$\mathbf{P}^2 = \begin{pmatrix} 0 & 1 \\ 0 & 0 \end{pmatrix}.$$

Step 7: Search the 0 row vector in $\mathbf{P}^2$. Obviously, the entries of the second row are all 0. This row is the third row in **P**. So x_3 is superior to x_1.
Step 8: By deleting the fourth row and the fourth column, the first row and the first column, and the third row and the third column in **P**, we get $\mathbf{P}^3 = (0)$.
Step 9: In the light of Theorem 3.3 and the previous corresponding algorithm, the FNPR has the property of satisfactory transitivity; and we find the most superior alternative x_2.

Therefore, the priority chain of the alternatives set{ $X = \{x_1, x_2, x_3, x_4\}$ is given as follows:

$$x_2 \succ x_3 \succ x_1 \succ x_4.$$

3.3 Properties and Priorities of IFNCPRs with Multiplicative Consistency

3.3.1 Properties of IFNCPRs with Multiplicative Consistency

An IFNCPR $\mathbf{R}=(r_{ij})_{n\times n}$ and an IFNRPR $\mathbf{A}=(a_{ij})_{n\times n}$ can be transformed into each other from the point of the view of multiplicative consistency. This subsection is based on (Gong and Liu, 2006; Xiao, Zhong et al., 2002).

Theorem 3.6. An IFNCPR $\mathbf{R}=(r_{ij})_{n\times n}$ and an IFNRPR $\mathbf{A}=(a_{ij})_{n\times n}$ can be transformed into each other by employing the following formulas:

$$r_{ij}=(1+a_{ji})^{-1},\quad i,j\in \mathrm{N} \tag{3.13}$$

$$a_{ij}=r_{ji}^{-1}-1,\quad i,j\in \mathrm{N} \tag{3.14}$$

Proof. Let $\boldsymbol{R}=(r_{ij})_{n\times n}$ be an IFNCPR. According to Eq. (3.13), if $r_{ii}=[0.5,0.5]$, then $a_{ii}=[1,1]$. If $a_{ijl}=r_{ijl}/r_{jiu}$ and $a_{jiu}=r_{jiu}/r_{ijl}$, then $a_{ijl}a_{jiu}=1$.

Similarly, we can show $a_{jil}a_{iju}=1$.

Thus, $\boldsymbol{A}=(a_{ij})_{n\times n}$ is an IFNRPR.

Let $\boldsymbol{A}=(a_{ij})_{n\times n}$ be an IFNRPR. According to Eq. (3.14), it follows that if $a_{ii}=[1,1]$, then $r_{ii}=[0.5,0.5]$. Meanwhile, $r_{ijl}+r_{jiu}=\frac{1}{1+a_{jiu}}+\frac{1}{1+a_{ijl}}=\frac{1}{1+a_{jiu}}+\frac{1}{1+\frac{1}{a_{jiu}}}=\frac{1+a_{jiu}}{1+a_{jiu}}=1$.

Similarly, we can show $a_{jil}+a_{iju}=1$.

Thus, $\boldsymbol{R}=(r_{ij})_{n\times n}$ is a IFNCPR. QED.

In the following, we will establish the concept of multiplicatively consistent IFNCPRs.

Let $\boldsymbol{V}=(v_1\ v_2\ \cdots\ v_n)^T$ be the priority vector of a multiplicatively consistent IFNRPR. Then, we have $a_{ij}=v_i v_j^{-1}, i,j\in N$. Let $r_{ij}=(1+a_{ji})^{-1}$, that is,

$$r_{ij}=\frac{1}{1+\frac{v_j}{v_i}}=\frac{1}{1+\frac{[v_{jl},v_{ju}]}{[v_{il},v_{iu}]}}=[\frac{v_{il}}{v_{il}+v_{ju}},\frac{v_{iu}}{v_{jl}+v_{iu}}].$$

If we let $\boldsymbol{\omega}=(\omega_1\ \omega_2\ \cdots\ \omega_n)^T$ be the priority vector of $\boldsymbol{R}=(r_{ij})_{n\times n}$, and

$$r_{ij}=[r_{ijl},r_{iju}]=\frac{1}{1+\frac{\omega_j}{\omega_i}}=[\frac{\omega_{il}}{\omega_{il}+\omega_{ju}},\frac{\omega_{iu}}{\omega_{jl}+\omega_{iu}}] \tag{3.15}$$

then we have

$$\frac{1}{r_{ij}}-1=[\frac{1}{r_{iju}}-1,\frac{1}{r_{ijl}}-1]=[\frac{\omega_{jl}}{\omega_{iu}},\frac{\omega_{ju}}{\omega_{il}}]$$

$$\frac{1}{r_{jk}}-1=[\frac{1}{r_{jku}}-1,\frac{1}{r_{jkl}}-1]=[\frac{\omega_{kl}}{\omega_{ju}},\frac{\omega_{ku}}{\omega_{jl}}]$$

$$\frac{1}{r_{ki}}-1=[\frac{1}{r_{kiu}}-1,\frac{1}{r_{kil}}-1]=[\frac{\omega_{il}}{\omega_{ku}},\frac{\omega_{iu}}{\omega_{kl}}]$$

Therefore, we have that

$$[\frac{1}{r_{iju}}-1][\frac{1}{r_{jku}}-1][\frac{1}{r_{kiu}}-1]=\frac{\omega_{jl}}{\omega_{iu}}\frac{\omega_{kl}}{\omega_{ju}}\frac{\omega_{il}}{\omega_{ku}}=\frac{\omega_{il}}{\omega_{ju}}\frac{\omega_{jl}}{\omega_{ku}}\frac{\omega_{kl}}{\omega_{iu}}=[\frac{1}{r_{jiu}}-1][\frac{1}{r_{kju}}-1][\frac{1}{r_{iku}}-1] \tag{3.16}$$

Similarly, we have

$$[\frac{1}{r_{ijl}}-1][\frac{1}{r_{jkl}}-1][\frac{1}{r_{kil}}-1]=[\frac{1}{r_{jil}}-1][\frac{1}{r_{kjl}}-1][\frac{1}{r_{ikl}}-1] \tag{3.17}$$

That is,

$$[\frac{1}{r_{ij}}-1][\frac{1}{r_{jk}}-1][\frac{1}{r_{ki}}-1]=[\frac{1}{r_{ji}}-1][\frac{1}{r_{kj}}-1][\frac{1}{r_{ik}}-1] \tag{3.18}$$

Thus we have the concept of multiplicatively consistent IFNCPRs as follows.

Definition 3.11. An IFNCPR $\boldsymbol{R}=(r_{ij})_{n\times n}$ is said to be multiplicatively consistent, if for all $i,j,k\in N$, Eq. (3.18) holds true. Any matrix satisfying the multiplicative consistency property is referred to as a multiplicatively consistent matrix.

As is well known, transitivity is an important issue which many scholars has been concerned with. A consistent preference must satisfy the properties of transitivity, such as the restricted max-max transitivity, the general transitivity, the weak transitivity, etc. In the following, we will generalize the transitivity of FPR (Tanino,1984; Herrera et al., 2004) to the case of IFNCPR. At the same time, we will construct the relationship between the property of multiplicative consistency and various transitivity properties of IFNCPRs in order to show that the previously defined concept of consistency is reasonable.

Definition 3.12. Let $\boldsymbol{R}=(r_{ij})_{n\times n}$ be a IFNCPR. For all $i,j,k\in N, i\neq j\neq k$ such that $r_{ij}\geq[\frac{1}{2},\frac{1}{2}], r_{jk}\geq[\frac{1}{2},\frac{1}{2}]$, if $r_{jkl}\geq r_{ijl}$ and $r_{jku}\geq r_{iju}$ imply $r_{ikl}\geq r_{jkl}$ or $r_{iku}\geq r_{jku}$; and $r_{ijl}\geq r_{jkl}$ and $r_{iju}\geq r_{jku}$ imply $r_{ikl}\geq r_{ijl}$ or $r_{iku}\geq r_{iju}$, then $\boldsymbol{R}$ is said to have the property of restricted max-max transitivity.

Definition 3.12 implies that if the degree to which a decision alternative X_i is preferred to another decision alternative X_j is the membership interval r_{ij}, and the degree to which the decision alternative X_j is preferred to a third decision alternative X_k is the membership interval r_{jk}, then the degree to which the alternative X_i is preferred to the alternative X_k is at least the lower limit of the membership interval r_{jk} or the upper limit of the membership interval r_{jk}.

Theorem 3.7. Each multiplicatively consistent IFNCPR $\boldsymbol{R} = (r_{ij})_{n\times n}$ has the property of restricted max-max transitivity.

Proof. Assume $r_{ij} \geq [\frac{1}{2},\frac{1}{2}]$, $r_{jk} \geq [\frac{1}{2},,\frac{1}{2}]$. We only need to prove that if $r_{jkl} \geq r_{ijl}$ and $r_{jku} \geq r_{iju}$, then $r_{ikl} \geq r_{jkl}$ or $r_{iku} \geq r_{jku}$.

Suppose for the purpose of producing a contradiction that there exist $i_0, j_0, k_0 \in N, i_0 \neq j_0 \neq k_0$ such that if $r_{j_0k_0l} \geq r_{i_0j_0l}$ and $r_{j_0k_0u} \geq r_{i_0j_0u}$, then $r_{i_0k_0l} < r_{j_0k_0l}, r_{i_0k_0u} < r_{j_0k_0u}$. Obviously, $1-r_{i_0k_0l} > 1-r_{j_0k_0l}$. That is, $r_{k_0i_0u} > r_{k_0j_0u}$. So we have

$$\frac{1}{r_{i_0k_0u}}-1 > \frac{1}{r_{j_0k_0u}}-1, \frac{1}{r_{k_0j_0u}}-1 > \frac{1}{r_{k_0i_0u}}-1 \tag{3.19}$$

Since $r_{i_0j_0} \geq [\frac{1}{2},\frac{1}{2}]$ implies $r_{j_0i_0} \leq [\frac{1}{2},\frac{1}{2}]$, we have

$$0 < \frac{1}{r_{i_0j_0u}}-1 \leq 1, \frac{1}{r_{j_0i\,u}}-1 \geq 1 \tag{3.20}$$

By applying Eqs. (3.16) and (3.20), we have

$$\begin{aligned}
&(\frac{1}{r_{j_0k_0u}}-1)(\frac{1}{r_{k_0i_0u}}-1) \geq (\frac{1}{r_{i_0j_0u}}-1))(\frac{1}{r_{j_0k_0u}}-1)(\frac{1}{r_{k_0i_0u}}-1)\\
&= (\frac{1}{r_{j_0i_0u}}-1)(\frac{1}{r_{k_0j_0u}}-1)(\frac{1}{r_{i_0k_0u}}-1) \geq (\frac{1}{r_{k_0j_0u}}-1)(\frac{1}{r_{i_0k_0u}}-1)
\end{aligned}$$

Meanwhile, from Eq. (3.19), we have

$$(\frac{1}{r_{j_0k_0u}}-1)(\frac{1}{r_{k_0i_0u}}-1) < (\frac{1}{r_{k_0j_0u}}-1)(\frac{1}{r_{i_0k\,u}}-1) \tag{3.21}$$

which contradicts Eq. (21). That is, the condition $r_{ikl} \geq r_{jkl}$ or $r_{iku} \geq r_{jku}$ holds for all $i, j, k \in N, i \neq j \neq k$. QED.

Theorem 3.7 means that for all $i, j, k \in N, i \neq j \neq k$, the upper limit degree of the membership satisfies $r_{iku} \geq \max\{r_{iju}, r_{jku}\}$, and the lower limit degree of the membership $r_{ikl} \geq \max\{r_{ijl}, r_{jkl}\}$.

Definition 3.13. Let $\boldsymbol{R} = (r_{ij})_{n\times n}$ be an IFNCPR. For all $i, j, k \in N$, $i \neq j \neq k$, if

When $0.5 \leq \lambda \leq 1$, $r_{ij} \geq [\lambda, \lambda], r_{jk} \geq [\lambda, \lambda]$ imply $r_{iku} \geq \lambda$; and
When $0 < \lambda \leq 0.5$, $r_{ij} \leq [\lambda, \lambda], r_{jk} \leq [\lambda, \lambda]$ imply $r_{ikl} \leq \lambda$.

then $\boldsymbol{R}$ is said to have the property of general consistency.

Notice that we attempt to discuss the property of the general consistency of IFNCPRs from different angles: The condition of the general consistency of IFNCPRs in this section is stronger than that in previous Section 3.2.

Definition 3.1.13 implies that if the degree, with which a decision alternative X_i is preferred to another decision alternative X_j, satisfies that the lower limit $r_{ij} \geq [\lambda, \lambda]$, and the degree, with which the decision alternative X_j is preferred to X_k, satisfies that the lower limit $r_{jk} \geq [\lambda, \lambda]$, then the degree, with which the alternative X_i is preferred to the alternative X_k, satisfies at least the upper limit $r_{iku} \geq \lambda$.

Theorem 3.8. Each multiplicatively consistent IFNCPR has the general consistency.

Proof. Let $\boldsymbol{R} = (r_{ij})_{n\times n}$ be a multiplicatively consistent IFNP. For any $i, j, k \in \mathrm{N}$, satisfying $i \neq j \neq k$. When $0.5 \leq \lambda \leq 1$, if $r_{ij} \geq [\lambda, \lambda], r_{jk} \geq [\lambda, \lambda]$, Theorem 3.7 then implies that $r_{iku} \geq \lambda$. When $0 < \lambda \leq 0.5$, if $r_{ij} \leq [\lambda, \lambda],\ r_{jk} \leq [\lambda, \lambda]$, it follows that $r_{ji} \geq [1-\lambda, 1-\lambda],\ r_{kj} \geq [1-\lambda, 1-\lambda]$. In the following, we verify $r_{ikl} \leq \lambda$.

Suppose for the purpose of producing a contradiction that there exist $i_0, j_0, k_0 \in N, i_0 \neq j_0 \neq k_0$, such that when $r_{i_0k_0l} > \lambda$, $r_{i_0k_0u} > \lambda$. That is, $r_{i_0k_0} > [\lambda, \lambda]$, and $r_{k_0i_0} < [1-\lambda, 1-\lambda]$. Obviously,

$$\frac{1}{r_{i_0j_0l}} - 1 \geq \frac{1}{\lambda} - 1, \quad \frac{1}{r_{j_0k_0l}} - 1 \geq \frac{1}{\lambda} - 1, \quad \frac{1}{r_{j_0i_0l}} - 1 \leq \frac{1}{1-\lambda} - 1, \quad \frac{1}{r_{k_0j_0l}} - 1 \leq \frac{1}{1-\lambda} - 1 \tag{3.22}$$

From Eqs. (3.17) and (3.22), we have

$$\begin{aligned}
&(\frac{\lambda}{1-\lambda})(\frac{\lambda}{1-\lambda})(\frac{1}{r_{i_0k_0l}} - 1) \geq (\frac{1}{r_{j_0i_0l}} - 1)(\frac{1}{r_{k_0j_0l}} - 1)(\frac{1}{r_{i_0k_0l}} - 1) \\
&= (\frac{1}{r_{i_0j_0l}} - 1)(\frac{1}{r_{j_0k_0l}} - 1)(\frac{1}{r_{k_0i_0l}} - 1) \geq (\frac{1-\lambda}{\lambda})(\frac{1-\lambda}{\lambda})(\frac{1}{r_{k_0i_0l}} - 1)
\end{aligned} \tag{3.23}$$

That is,

$$r_{ikl} < \frac{\lambda^3}{\lambda^3 + (1-\lambda)^3}$$

It easily to see that when $0 < \lambda \leq 0.5$, the following inequality

$$\frac{\lambda^3}{\lambda^3 + (1-\lambda)^3} \leq \lambda$$

holds true. That is, we have

$$r_{ikl} < \frac{\lambda^3}{\lambda^3 + (1-\lambda)^3} \leq \lambda < r_{ikl}$$

which is a contradiction. Thus we have $r_{ikl} \leq \lambda$. QED.

Theorem 3.8 shows that when $0.5 \leq \lambda \leq 1$, if $r_{ij} \geq [\lambda, \lambda]$, $r_{jk} \geq [\lambda, \lambda]$, then $r_{iku} \geq \lambda$. That is, the uppoer limit r_{iku} of the interval r_{ik} is bigger than λ; and that when $0 < \lambda \leq 0.5$, if $r_{ij} \leq [\lambda, \lambda], r_{jk} \leq [\lambda, \lambda]$, then the lower limit r_{ikl} of the interval r_{ik} is smaller than λ.

Definition 3.14. Let $\boldsymbol{R} = (r_{ij})_{n\times n}$ be an INFNPR. For all $i, j, k \in$ N, satisfying $i \neq j \neq k$, if

$$r_{ij} \geq [\frac{1}{2}, \frac{1}{2}],\ r_{jk} \geq [\frac{1}{2}, \frac{1}{2}] \Rightarrow r_{ik}^{+} \geq \frac{1}{2};$$

or

$$r_{ij} \leq [\frac{1}{2}, \frac{1}{2}],\ r_{jk} \leq [\frac{1}{2}, \frac{1}{2}] \Rightarrow r_{ik}^{-} \leq \frac{1}{2}$$

then $\boldsymbol{R}$ is said to have the weak consistency.

Notice that we now attempt to discuss the property of weak consistency of IFNCPRs in different lights: The condition of the property of weak consistency of IFNCPRs in this section is stronger than that in the previous Section 3.2.

If we let $\lambda = \frac{1}{2}$, then Corollary 3.3 below follows immediately from Theorem3.8.

Corollary 3.3. Each multiplicatively consistent IFNCPR has the property of weak consistency.

3.3.2 *Priorities of IFNCPRs with Multiplicative Consistency*

Let $\boldsymbol{\omega}=(\omega_1\ \omega_2\ \cdots\ \omega_n)^T$ be the priority vector of the INFPR $\boldsymbol{R}=(r_{ij})_{n\times n}$, where $\omega_i=[\omega_{il},\omega_{iu}], i\in$ N. If $\boldsymbol{R}=(r_{ij})_{n\times n}$ is multiplicatively consistent, then for all $i,j\in$ N, Eq. (3.15) holds true. That is,

$$\begin{cases} r_{ijl}(\omega_{il}+\omega_{ju})=\omega_{il} \\ r_{iju}(\omega_{iu}+\omega_{jl})=\omega_{iu} \end{cases} \tag{3.24}$$

In a real decision making situation, it is hard for the decision makers to provide consistent INFPRs. That means that Eq. (3.24) does not hold true. So we introduce the following deviation function

$$\begin{cases} g_{ijl}=[r_{ijl}(\omega_{il}+\omega_{ju})-\omega_{il}]^2 \\ g_{iju}=[r_{iju}(\omega_{iu}+\omega_{jl})-\omega_{iu}]^2 \end{cases} \tag{3.25}$$

It is clear that the smaller the deviation function value is, the better consistency of the decision maker's judgment. We construct the following optimization model 3.1:

$$\begin{aligned} &min \quad g_{ijl}=[r_{ijl}(\omega_{il}+\omega_{ju})-\omega_{il}]^2 \\ &min \quad g_{iju}=[r_{iju}(\omega_{iu}+\omega_{jl})-\omega_{iu}]^2 \\ &s.t. \quad \begin{cases} 0\le\omega_{il}\le\omega_{iu}\le 1 \\ 0\le\sum_{i=1}^{n}\omega_{il}\le 1\le\sum_{i=1}^{n}\omega_{iu}, i,j\in N \end{cases} \end{aligned}$$

There is no preference between two deviation function values g_{ijl} and g_{iju}, for $i,j\in N$. We construct the following nonlinear programming model 3.2:

$$\begin{aligned} &min \quad J=\sum_{j=1}^{n}\sum_{i=1}^{n}[r_{ijl}(\omega_{il}+\omega_{ju})-\omega_{il}]^2+[r_{iju}(\omega_{iu}+\omega_{jl})-\omega_{iu}]^2 \\ &s.t. \quad \begin{cases} 0\le\sum_{i=1}^{n}\omega_{il}\le 1\le\sum_{i=1}^{n}\omega_{iu} \\ 0\le\omega_{il}\le\omega_{iu}\le 1, \quad i\in N \end{cases} \end{aligned}$$

Similarly, we can also introduce the following deviation function

$$\begin{cases} g_{ijl}=|\, r_{ijl}(\omega_{il}+\omega_{ju})-\omega_{il}\,| \\ g_{iju}=|\, r_{iju}(\omega_{iu}+\omega_{jl})-\omega_{iu}\,| \end{cases} \tag{3.26}$$

Similar to that of Model 3.2, we can also construct the following nonlinear programming model 3.3:

$$min \quad J = \sum_{j=1}^{n} \sum_{i=1}^{n} | r_{ijl} (\omega_{il} + \omega_{ju}) - \omega_{il} | + | r_{iju} (\omega_{iu} + \omega_{jl}) - \omega_{iu} |$$

$$s.t. \begin{cases} 0 \le \sum_{i=1}^{n} \omega_{il} \le 1 \le \sum_{i=1}^{n} \omega_{iu} \\ 0 \le \omega_{il} \le \omega_{iu} \le 1, \quad i, j \in N \end{cases}$$

3.3.3 A Numerical Example

Consider a set $\{X_1, X_2, X_3, X_4\}$ of decision alternatives. Assume that the IFNCPR given by the decision maker is as follows

$$\begin{pmatrix} [0.5,0.5] & [0.2,0.4] & [0.3,0.6] & [0.5,0.7] \\ [0.6,0.8] & [0.5,0.5] & [0.7,0.9] & [0.6,0.8] \\ [0.4,0.7] & [0.1,0.3] & [0.5,0.5] & [0.7,0.8] \\ [0.3,0.5] & [0.2,0.4] & [0.2,0.3] & [0.5,0.5] \end{pmatrix}$$

Utilizing the 'Matlab Optimization Toolbox', we produce the solution to Model 3.3 as follows:

$$\omega_1 = [0.1234, 0.2147],\ \omega_2 = [0.3619, 0.4583],$$
$$\omega_3 = [0.1120, 0.1901],\ \omega_4 = [0.0906, 0.1370].$$

The ranking method of intervals is introduced (Xu and Da, 2003b) as follows:

$$p_{ij} = P(x_i \ge x_j) = max\{1 - max(\frac{\omega_{ju} - \omega_{il}}{l_{x_i} + l_{x_j}}, 0), 0\} \tag{3.27}$$

$$v_i = \frac{1}{n(n-1)} (\sum_{j=1}^{n} p_{ij} + \frac{n}{2} - 1),\ i, j \in N \tag{3.27a}$$

where $l_{x_i} = \omega_{iu} - \omega_{il}, i \in N$, $\boldsymbol{P} = (p_{ij})_{n \times n}$ is the possibility degree matrix, and $\boldsymbol{V} = (v_1, v_2, v_3, v_4)$ the priority vector of $\boldsymbol{P}$. Obviously, we have

$$v_1 = 0.2506, v_2 = 0.3750, v_3 = 0.2244, v_4 = 0.1499.$$

The ranking of the intervals ω_i, i = 1, 2, 3, 4, is $\omega_2 \underset{1}{\ge} \omega_1 \underset{0.6062}{\ge} \omega_3 \underset{0.7992}{\ge} \omega_4$, where p_{21} = 1, p_{13} = 0.6062, p_{34} = 0.7992. The order of optimal ranking of the decision alternatives is given by

$$X_2 \underset{1}{\succ} X_1 \underset{0.6062}{\succ} X_3 \underset{0.7992}{\succ} X_4.$$

3.4 Group Decision Making Based on Complementary Judgment Matrices of Incomplete Interval Fuzzy Numbers

Let $\{X_1, X_2, \cdots, X_n\}$ be a finite set of decision alternatives. In order to rank these alternatives from the best to the worst, the decision makers usually construct their interval preference relations (complementary judgment matrices of interval fuzzy numbers) to express their judgment information of pairwise comparisons. If some entries of a preference relation are not present, we say that the preference relation is incomplete. Only when there exist direct or indirect comparisons of every two decision alternatives, we can utilize this finite information to rank all the decision alternatives (Saaty, 1980; Xu, 2004b).

3.4.1 Least Squares Priority Model of Complementary Preference Relations of Incomplete Interval Fuzzy Numbers

Let us suppose that the pairwise comparisons of n decision alternatives $X_1, X_2, \cdots, X_n$ are given by m decision makers. (This subsection is based on (Zhang, Yu et al., 2006), for more details, please consult with this reference). For simplicity, let $N = \{1, 2, \cdots, n\}$ and $M = \{1, 2, \cdots, m\}$. Consider the following complementary judgment matrix of interval fuzzy numbers:

$$\boldsymbol{R}_s = \begin{pmatrix} r_{11s} & r_{12s} & \cdots & r_{1ns} \\ r_{21s} & r_{22s} & \cdots & r_{21ns} \\ \cdots & \cdots & \cdots & \cdots \\ r_{n1s} & r_{n2s} & \cdots & r_{nns} \end{pmatrix}, s \in M$$

where the (i,j) entry r_{ijs} denotes the degree of pairwise preference between the decision alternative X_i over X_j as estimated by the s th decision maker, and takes the form of interval fuzzy numbers, for all $i, j \in N, s \in M$.

In many practical cases, the decision makers may possibly provide incomplete information. Therefore, for $i_0, j_0 \in N$, $s_0 \in M$, $r_{i_0 j_0 s_0}$ may be empty. In this case, we denote $r_{i_0 j_0 s_0} = -$, where we use d_{ij} to denote that there are d_{ij} pairwise comparisons between the decision alternatives X_i and X_j. Obviously, $0 \le d_{ij} \le m$.

Let $W = (w_1, w_2 \cdots, w_n)^T$ be a priority vector of the decision alternatives $X_1, X_2, \cdots, X_n$ derived by using the reciprocal preference relation $\boldsymbol{A}_s = (a_{ijs})_{n \times n}$ of interval fuzzy numbers with complete consistency. Then $a_{ij} = w_i / w_j$, where

$w_i, i \in N$, is a positive interval fuzzy number. From Eq. (3.13), it follows that if we let $V = (v_1, v_2 \cdots, v_n)^T$ be a priority vector for the complementary judgment matrix $R_s = (r_{ijs})_{n \times n}$ of interval fuzzy numbers with complete consistency, then

$$r_{ijs} = 0.5 + 0.2log_3^{v_i/v_j} \tag{3.28}$$

where $v_i, i \in N$, is a positive interval fuzzy number.

However, in the general case, Eq. (3.29) cannot hold true for inconsistent estimations. For such cases, we can obtain an optimal priority $V = (v_1, v_2 \cdots, v_n)^T$ by minimizing the squared error between a_{ijs} and $0.5 + 0.2log_3^{v_i/v_j}$ for all $i, j \in N, s \in M$, where we assume that all the decision makers are of the same importance, as follows:

$$min \sum_{i=1}^{n} \sum_{j=1}^{n} \sum_{s=1}^{d_{ij}} (a_{ijs} - 0.5 - 0.2log_3^{v_i/v_j})^2 \tag{3.29}$$

Let $y_{ijs} = a_{ijs} - 0.5$, $x_i = 0.2log_3^{v_i}$. Then we have

$$min \sum_{i=1}^{n} \sum_{j=1}^{n} \sum_{s=1}^{d_{ij}} (y_{ijs} - x_i + x_j)^2 \tag{3.30}$$

The optimal solution to Eqs. (3.30) is given as follows:

$$x_i \sum_{j=1, j \neq i}^{n} d_{ij} - \sum_{j=1, j \neq i}^{n} d_{ij} x_j = \sum_{j=1, j \neq i}^{n} \sum_{s=1}^{d_{ij}} y_{ijs}, i \in N \tag{3.31}$$

where $y_{ijs} = (l_{ijs}, u_{ijs})$, $x_i = (l_i, u_i), i, j \in N, s \in M$, are positive interval fuzzy numbers. Eqs. (3.30) and (3.31) can be transformed into the following equations:

$$l_i \sum_{j=1, j \neq i}^{n} d_{ij} - \sum_{j=1, j \neq i}^{n} d_{ij} u_j = \sum_{j=1, j \neq i}^{n} \sum_{s=1}^{d_{ij}} l_{ijs}, i \in N \tag{3.32}$$

$$u_i \sum_{j=1, j \neq i}^{n} d_{ij} - \sum_{j=1, j \neq i}^{n} d_{ij} l_j = \sum_{j=1, j \neq i}^{n} \sum_{s=1}^{d_{ij}} u_{ijs}, i \in N \tag{3.33}$$

For simplicity, let $\sum_{j=1, j \neq i}^{n} \sum_{s=1}^{d_{ij}} l_{ijs} = b_i$ and $\sum_{j=1, j \neq i}^{n} \sum_{s=1}^{d_{ij}} u_{ijs} = c_i$, $i \in N$. The matrix format of Eqs. (3.32) and (3.33) is

$$Eh = d \tag{3.34}$$

where

$$E=\begin{pmatrix} S & T \\ T & S \end{pmatrix}, T=\begin{pmatrix} 0 & -d_{12} & \cdots & -d_{1n} \\ -d_{21} & 0 & \cdots & -d_{2n} \\ \cdots & \cdots & \cdots & \cdots \\ -d_{n1} & -d_{n2} & \cdots & 0 \end{pmatrix}, S=\begin{pmatrix} \sum_{j=1,j\neq 1}^{n} d_{1j} & 0 & \cdots & 0 \\ 0 & \sum_{j=1,j\neq 2}^{n} d_{2j} & \cdots & 0 \\ \cdots & \cdots & \cdots & \cdots \\ 0 & 0 & \cdots & \sum_{j=1,j\neq n}^{n} d_{nj} \end{pmatrix},$$

$$h=(l_1,l_2,\cdots,l_n,u_1,u_2,\cdots,u_n)^T,\ d=(b_1,b_2,\cdots,b_n,c_1,c_2,\cdots,c_n)^T.$$

By the complementary property of preference relations, we have $\sum_{i=1}^{n}(b_i+c_i)=0$.

In the following, we will discuss the condition of existence for Eq. (3.34), and then provide the general solution. Firstly, let us consider the properties of matrix E.

Property 1. E is a symmetric matrix.

Property 2. E is a semi-positive matrix.

The conclusion of Property 1 is obvious so that we will only provide the proof of Property 2.

Secondly, we introduce an important lemma.

Lemma 3.1 (*Geršgorin* [65]). Let $A=(a_{ij})_{n\times n}\in M_n$. And let

$$R_i'(A)=\sum_{j=1,j\neq i}^{n}|a_{ij}|, 1\le i\le n$$

denote the deleted absolute row sums of A. Then all the eigenvalues of the matrix A are located in the union of n discs:

$$\bigcup_{i=1}^{n}\{z\in C:|z-a_{ii}|\le R_i'(A)\}$$

where C denotes field of complex numbers.

By employing Lemma 3.1, we now look at the proof of Property 2.

Proof of Property 2. For $E=(e_{ij})_{n\times n}$, we have $R_i'(E)=\sum_{j=1,j\neq i}^{n}d_{ij}, 1\le i\le 2n$. Since $e_{ii}=\sum_{j=1,j\neq i}^{n}d_{ij}$, from Lemma 1, we have $|z-e_{ii}|\le R_i'(E)\Rightarrow 0\le z\le 2\sum_{j=1,j\neq i}^{n}d_{ij}$. That is,

all the eigenvalues of E are nonnegative. Since $E^T = E$, E is semipositive definite.
QED.

Now let us consider the solution to Eq. (3.34).

Theorem 3.9. If $R(E) = R(E,d)$, then the equation $Eh = d$ must have solutions, in which the general solution is given as follows:

$$h = E^+ d + (I - E^+ E)x \tag{3.35}$$

where E^+ is the generalized pseudo-inverse of E, I the identity matrix of order $2n$, and $x = \{x_1, x_2, \cdots, x_{2n}\}^T$ an arbitrary column vector.

The proof follows from (Gong, Bai et el., 2009).

In particular, when $R(E) = R(E,d) = 2n-1$, because E is semipositive definite, there exists an orthogonal matrix P (Horn and Johnson, 1985) such that $E^+ = P \wedge^+ P^T$, where $\wedge$ is a diagonal matrix with the eigenvalues $\lambda_1, \cdots, \lambda_{2n-1}, 0$ of E located along the diagonal, and

$$I - E^+ E = \begin{pmatrix} \frac{1}{2n} & \frac{1}{2n} & \cdots & \frac{1}{2n} \\ \frac{1}{2n} & \frac{1}{2n} & \cdots & \frac{1}{2n} \\ \cdots & \cdots & \cdots & \cdots \\ \frac{1}{2n} & \frac{1}{2n} & \cdots & \frac{1}{2n} \end{pmatrix} = T$$

Thus the solution to $Eh = d$ is given below:

$$\begin{aligned} h &= E^+ d + Tx \\ &= E^+ d + \left(\frac{1}{2n}\sum_{i=1}^{2n} x_i, \frac{1}{2n}\sum_{i=1}^{2n} x_i, \cdots, \frac{1}{2n}\sum_{i=1}^{2n} x_i\right)^T. \end{aligned}$$

where $x = (x_1, x_2, \cdots, x_{2n})^T$ is an arbitrary real vector. If we let $\frac{1}{2n}\sum_{i=1}^{2n} x_i = p$, then p is also an arbitrary real number.

Let $P = (p, p, \cdots, p)^T$. Then the solution to the equation $Eh = d$ can be written in the following more general form:

$$h = E^+ d + P \tag{3.36}$$

In view of Theorem 4.1, we can readily obtain the following corollary:

Corollary 3.4. If $R(E) = R(E,d) = 2n-1$, then the equation $Eh = d$ must have solutions, of which the general solution is $h = E^+ d + P$, where E^+ is the generalized pseudoinverse of E, $P=(p,p,\cdots,p)^T$ an arbitrary column vector of $2n$ dimension. When $p=0$, the particular solution to $Eh=d$ is

$$h=E^{+}d \tag{3.37}$$

For $x_i=0.2log_3^{v_i}$, we have $v_i=243^{x_i}, i\in N$. Hence the general solution to Eq. (3.31) is $v_i=(243^{l_i}, 243^{u_i})$.

3.4.2 A Numerical Example

Supposed that the complementary judgment matrices of incomplete interval fuzzy numbers, as presented by three decision makers d_1,d_2,d_3 on a set $X=\{X_1,X_2,X_3,X_4\}$ of decision alternatives, are respectively as follows, where it is also assumed that each decision maker has the same importance:

$$A_1=\begin{pmatrix}[0.5,0.5] & - & [0.6,0.7] & - \\ - & [0.5,0.5] & - & [0.6,0.7] \\ [0.3,0.4] & - & [0.5,0.5] & - \\ - & [0.3,0.4] & - & [0.5,0.5]\end{pmatrix}$$

$$A_2=\begin{pmatrix}[0.5,0.5] & - & - & [0.7,0.8] \\ - & [0.5,0.5] & [0.8,0.9] & - \\ - & [0.1,0.2] & [0.5,0.5] & - \\ [0.2,0.3] & - & - & [0.5,0.5]\end{pmatrix}$$

$$A_3=\begin{pmatrix}[0.5,0.5] & - & [0.7,0.8] & [0.5,0.6] \\ - & [0.5,0.5] & [0.6,0.7] & - \\ [0.2,0.3] & [0.3,0.4] & [0.5,0.5] & - \\ [0.4,0.5] & - & - & [0.5,0.5]\end{pmatrix}$$

Step 1: Construct the equation $Eh=d$, where $d=(0.5,0.5,-1.1,-0.6,0.9,0.8,-0.7,-0.3)^T$, and

$$E=\begin{pmatrix}4 & 0 & 0 & 0 & 0 & 0 & -2 & -2 \\ 0 & 3 & 0 & 0 & 0 & 0 & -2 & -1 \\ 0 & 0 & 4 & 0 & -2 & -2 & 0 & 0 \\ 0 & 0 & 0 & 4 & -2 & -1 & -1 & 0 \\ 0 & 0 & -2 & -2 & 4 & 0 & 0 & 0 \\ 0 & 0 & -2 & -1 & 0 & 3 & 0 & 0 \\ -2 & -2 & 0 & 0 & 0 & 0 & 4 & 0 \\ -2 & -1 & -1 & 0 & 0 & 0 & 0 & 4\end{pmatrix}$$

Step 2: Test the rank of matrices E and (E,d). Obviously we have $R(E)=R(E,d)=7$. From Corollary 3.4, it follows that the equation $Eh=d$ must have at least one solution, while the general solution is $h=E^{+}d+P$. Let $P=0$, then

$$h=E^{+}d=(0.0305,0.0649,-0.1234,-0.0720,$$
$$0.1305,0.1649,-0.1234,-0.0720)^{T}.$$

where

$$E^{+}=\begin{pmatrix} U & V \\ V & U \end{pmatrix},$$

$$U=\begin{pmatrix} 0.4582 & 0.1954 & -0.2678 & -0.2533 \\ 0.2167 & 0.5507 & -0.3188 & -0.2852 \\ -0.3486 & -0.3658 & 0.4623 & 0.2030 \\ -0.1764 & -0.1803 & 0.0743 & 0.3355 \end{pmatrix},$$

and

$$V=\begin{pmatrix} -0.2918 & -0.3046 & 0.2322 & 0.2467 \\ -0.3333 & -0.3493 & 0.2812 & 0.2148 \\ 0.3014 & 0.3342 & -0.3377 & -0.2970 \\ 0.1736 & 0.1197 & -0.1257 & -0.1645 \end{pmatrix}.$$

Therefore, the priority vector of the group decision making is

$$(v_1,v_2,v_3,v_4)^{T} = ([1.1824,2.0480],[1.4283,2.4739],[0.5077,0.5077],[0.6733,0.6733])^{T}.$$

According to the ranking method of interval numbers, as proposed by (Xu and Da, 2003b), we derived the priority chain

$$X_2 \succ X_1 \succ X_4 \succ X_3.$$

3.5 Logarithmic Least Squares Priority Model of Collective IFNCPRs

3.5.1 Logarithmic Least Squares Priority Model of Collective IFNCPRs

An IFNCPR $\tilde{R}=(\tilde{r}_{ijs})_{n\times n}=([\mu_{ijs},p_{ijs}])_{n\times n}$ is multiplicatively consistent if there exists a priority vector $\Omega=(\omega_1\ldots\omega_n)^{T}=([\omega_{1l},\omega_{1u}]\ldots\ [\omega_{nl},\omega_{nu}])^{T}$ such that

$$\tilde{r}_{ijs} = 1/(1+\omega_j/\omega_i) = [\omega_{il}/(\omega_{il}+\omega_{ju}), \omega_{iu}/(\omega_{jl}+\omega_{iu})] \forall i, j \in N,$$

where $\omega_i = [\omega_{il}, \omega_{iu}] \subset [0,1]$. This subsection is based on (Gong and Cui, 2008), with which the reader is advised to consult.

Let $X = \{x_1, x_2, \ldots, x_n\}$ be a set of decision alternatives and $d = \{d_1, d_2, \ldots, d_m\}$ a set of decision makers. The preferences of the decision makers on the set X are respectively described by the IFNCPRs as follows:

$$\tilde{R} = (\tilde{r}_{ijs})_{n\times n} = ([\mu_{ijs}, p_{ijs}])_{n\times n} =$$

$$= \begin{pmatrix} [0.5,0.5] & \begin{Bmatrix} [\mu_{121}, p_{121}] \\ \vdots \\ [\mu_{12\delta_{12}}, p_{12\delta_{12}}] \end{Bmatrix} & \cdots & \begin{Bmatrix} [\mu_{1n1}, p_{1n1}] \\ \vdots \\ [\mu_{1n\delta_{1n}}, p_{1n\delta_{1n}}] \end{Bmatrix} \\ \begin{Bmatrix} [\mu_{211}, p_{211}] \\ \vdots \\ [\mu_{21\delta_{21}}, p_{21\delta_{21}}] \end{Bmatrix} & [0.5,0.5] & \cdots & \begin{Bmatrix} [\mu_{2n1}, p_{2n1}] \\ \vdots \\ [\mu_{2n\delta_{2n}}, p_{2n\delta_{2n}}] \end{Bmatrix} \\ \vdots & \vdots & \vdots & \vdots \\ \begin{Bmatrix} [\mu_{n11}, p_{n11}] \\ \vdots \\ [\mu_{n1\delta_{n1}}, p_{n1\delta_{n1}}] \end{Bmatrix} & \begin{Bmatrix} [\mu_{n21}, p_{n21}] \\ \vdots \\ [\mu_{n2\delta_{n2}}, p_{n2\delta_{n2}}] \end{Bmatrix} & \cdots & [0.5,0.5] \end{pmatrix}$$

where $\tilde{r}_{ijs} = [\mu_{ijs}, p_{ijs}]$ stand for the elements of the IFNCPRs $\tilde{R}$ satisfying that $\mu_{ijs} + p_{jis} = 1, \mu_{jis} + p_{ijs} = 1, s = 1, 2, \ldots, \delta_{ij}$, $\forall i, j \in N, i \neq j$, and δ_{ij}, $0 \le \delta_{ij} \le m$ denotes that there are δ_{ij} decision makers who presented their degrees of preference of the decision alternative x_i over x_j. It is clear that $\delta_{ij} = \delta_{ji}$. If there exist $i_0, j_0 \in N$ such that $0 < \delta_{i_0 j_0} < m$, then $m - \delta_{i_0 j_0}$ decision makers did not present their judgment information between the decision alternatives x_{i_0} and x_{j_0}; if there exist $i_0, j_0 \in N$ such that $\delta_{i_0 j_0} = 0$, then no decision maker presents the judgment information between the decision alternatives x_{i_0} and x_{j_0}. In this case, we write $\tilde{r}_{ijs} = -$. That means that the element $\tilde{r}_{i_0 j_0 s}$ in $\tilde{R}$ is absent and the IFNCPRs $\tilde{R}$ are called judgment matrices with incomplete information. For all $i, j \in N$, if $\delta_{ij} = m$, then all the decision makers present their judgment information regarding decision alternatives x_i and x_j, and $\tilde{R}$ stand for judgment matrices of complete information.

Let $\tilde{\eta} = ([\tilde{\omega}_{1l}, \tilde{\omega}_{1u}] \ldots [\tilde{\omega}_{il}, \tilde{\omega}_{iu}] \ldots [\tilde{\omega}_{nl}, \tilde{\omega}_{nu}])^T$ be the priority vector of the multiplicatively consistent IFNCPRs $\tilde{R}$, then we have

$$\mu_{ijs} = \frac{\tilde{\omega}_{il}}{\tilde{\omega}_{il} + \tilde{\omega}_{ju}}, \tag{3.38}$$

$$p_{ijs} = \frac{\tilde{\omega}_{iu}}{\tilde{\omega}_{jl} + \tilde{\omega}_{iu}}, \tag{3.39}$$

where $0 < \tilde{\omega}_{il} \leq \tilde{\omega}_{iu} \leq 1, i, j \in N$. Eqs. (3.38) and (3.39) are equivalent to the following equations:

$$ln(\frac{1}{\mu_{ijs}} - 1) = ln\tilde{\omega}_{ju} - ln\tilde{\omega}_{il}, \tag{3.40}$$

$$ln(\frac{1}{p_{ijs}} - 1) = ln\tilde{\omega}_{jl} - ln\tilde{\omega}_{iu}. \tag{3.41}$$

In reality, it is hard for a decision maker to be consistent and different decision makers may present different judgments. In consequence, we introduce the following deviation functions:

$$\tilde{\varepsilon}_{ij} = [ln(\frac{1}{\mu_{ijs}} - 1) + ln\tilde{\omega}_{il} - ln\tilde{\omega}_{ju}]^2, \tag{3.42}$$

$$\tilde{\gamma}_{ij} = [ln(\frac{1}{p_{ijs}} - 1) + ln\tilde{\omega}_{iu} - ln\tilde{\omega}_{jl}]^2. \tag{3.43}$$

It is clear that the smaller values the deviation functions take, the better the consistency of judgments is. In order to get the optimal priority vector of inconsistent IFNCPRs, let us introduce an optimization model as follows:

$$\min J = \sum_{i=1}^{n} \sum_{j=1, j \neq i}^{n} \sum_{s=1}^{\delta_{ij}} [ln(\frac{1}{\mu_{ijs}} - 1) + ln\tilde{\omega}_{il} - ln\tilde{\omega}_{ju}]^2 + [ln(\frac{1}{p_{ijs}} - 1) + ln\tilde{\omega}_{iu} - ln\tilde{\omega}_{jl}]^2$$

$$\text{s.t.} \quad \begin{cases} \tilde{\omega}_{il} + \sum_{j=1, j \neq i}^{n} \tilde{\omega}_{ju} \geq 1, & i \in N; \\ \tilde{\omega}_{iu} + \sum_{j=1, j \neq i}^{n} \tilde{\omega}_{jl} \leq 1, & i \in N; \\ \tilde{\omega}_{iu} - \tilde{\omega}_{il} \geq 0, & i \in N; \\ \tilde{\omega}_{il} > 0, & i \in N, \end{cases} \tag{3.44}$$

where the two constraints $\omega_{il} + \sum_{j=1, j\neq i}^{n} \omega_{ju} \geq 1$ and $\omega_{iu} + \sum_{j=1, j\neq i}^{n} \omega_{jl} \leq 1$ are the normalization constraints on the interval vector Ω (Gong, Bai et al., 2009).

Obviously, for all $i, j \in N$, if $\delta_{ij} = m$ or $\delta_{ij} = 1$, model (3.44) can be respectively regarded as a priority model for either a collective IFNCPR or an individual with complete information. For some $i_0, j_0 \in N$, if $0 < \delta_{i_0 j_0} < m$, we obtain a priority model for a collective IFNCPR with incomplete information.

3.5.2 A Numerical Example

Suppose that three decision makers respectively provide the following incomplete IFNCPRs $\{\tilde{R}_1, \tilde{R}_2, \tilde{R}_3\}$ on a set of four decision alternatives $X = \{x_1, x_2, x_3, x_4\}$.

$$\tilde{R}_1 = \begin{pmatrix} [0.5,0.5] & - & [0.6,0.7] & [0.8,0.9] \\ - & [0.5,0.5] & - & - \\ [0.3,0.4] & - & [0.5,0.5] & [0.7,0.8] \\ [0.1,0.2] & - & [0.2,0.3] & [0.5,0.5] \end{pmatrix};$$

$$\tilde{R}_2 = \begin{pmatrix} [0.5,0.5] & - & - & [0.6,0.8] \\ - & [0.5,0.5] & - & [0.1,0.3] \\ - & - & [0.5,0.5] & - \\ [0.2,0.4] & [0.7,0.9] & - & [0.5,0.5] \end{pmatrix};$$

$$\tilde{R}_3 = \begin{pmatrix} [0.5,0.5] & - & [0.7,0.8] & [0.8,0.9] \\ - & [0.5,0.5] & - & [0.1,0.1] \\ [0.2,0.3] & - & [0.5,0.5] & [0.8,0.9] \\ [0.1,0.2] & [0.9,0.9] & [0.1,0.2] & [0.5,0.5] \end{pmatrix}.$$

Step 1: Construct the optimal model as follows:

$$\begin{aligned}
\min J_2 = {} & (\ln(3/2)-\ln(\omega_{1l})+\ln(\omega_{3u}))^2+(\ln(3/7)-\ln(\omega_{3l})+\ln(\omega_{1u}))^2+ \\
& (\ln(7/3)-\ln(\omega_{1l})+\ln(\omega_{3u}))^2+(\ln(1/9)-\ln(\omega_{3l})+\ln(\omega_{1u}))^2+ \\
& (\ln(4/1)-\ln(\omega_{1l})+\ln(\omega_{4u}))^2+(\ln(1/9)-\ln(\omega_{4l})+\ln(\omega_{1u}))^2+ \\
& (\ln(3/2)-\ln(\omega_{1l})+\ln(\omega_{4u}))^2+(\ln(1/4)-\ln(\omega_{4l})+\ln(\omega_{1u}))^2+ \\
& (\ln(4/1)-\ln(\omega_{1l})+\ln(\omega_{4u}))^2+(\ln(1/9)-\ln(\omega_{4l})+\ln(\omega_{1u}))^2+ \\
& (\ln(1/9)-\ln(\omega_{2l})+\ln(\omega_{4u}))^2+(\ln(7/3)-\ln(\omega_{4l})+\ln(\omega_{2u}))^2+ \\
& (\ln(1/9)-\ln(\omega_{2l})+\ln(\omega_{4u}))^2+(\ln(9/1)-\ln(\omega_{4l})+\ln(\omega_{2u}))^2+ \\
& (\ln(3/7)-\ln(\omega_{3l})+\ln(\omega_{1u}))^2+(\ln(3/2)-\ln(\omega_{1l})+\ln(\omega_{3u}))^2+ \\
& (\ln(1/9)-\ln(\omega_{3l})+\ln(\omega_{1u}))^2+(\ln(7/3)-\ln(\omega_{1l})+\ln(\omega_{3u}))^2+ \\
& (\ln(7/3)-\ln(\omega_{3l})+\ln(\omega_{4u}))^2+(\ln(1/9)-\ln(\omega_{4l})+\ln(\omega_{3u}))^2+ \\
& (\ln(4/1)-\ln(\omega_{3l})+\ln(\omega_{4u}))^2+(\ln(1/9)-\ln(\omega_{4l})+\ln(\omega_{3u}))^2+ \\
& (\ln(1/9)-\ln(\omega_{4l})+\ln(\omega_{1u}))^2+(\ln(4/1)-\ln(\omega_{1l})+\ln(\omega_{4u}))^2+ \\
& (\ln(1/4)-\ln(\omega_{4l})+\ln(\omega_{1u}))^2+(\ln(2/3)-\ln(\omega_{1l})+\ln(\omega_{4u}))^2+ \\
& (\ln(1/9)-\ln(\omega_{4l})+\ln(\omega_{1u}))^2+(\ln(4/1)-\ln(\omega_{1l})+\ln(\omega_{4u}))^2+ \\
& (\ln(7/3)-\ln(\omega_{4l})+\ln(\omega_{2u}))^2+(\ln(1/9)-\ln(\omega_{2l})+\ln(\omega_{4u}))^2+ \\
& (\ln(9/1)-\ln(\omega_{4l})+\ln(\omega_{2u}))^2+(\ln(1/9)-\ln(\omega_{2l})+\ln(\omega_{4u}))^2+ \\
& (\ln(1/9)-\ln(\omega_{4l})+\ln(\omega_{3u}))^2+(\ln(7/3)-\ln(\omega_{3l})+\ln(\omega_{4u}))^2+ \\
& (\ln(9/1)-\ln(\omega_{4l})+\ln(\omega_{3u}))^2+(\ln(4/1)-\ln(\omega_{3l})+\ln(\omega_{4u}))^2
\end{aligned} \tag{3.45}$$

$$s.t.\quad \begin{cases}
\omega_{1l}+\omega_{2u}+\omega_{3u}+\omega_{4u}\ge 1,\ \omega_{2l}+\omega_{1u}+\omega_{3u}+\omega_{4u}\ge 1, \\
\omega_{3l}+\omega_{1u}+\omega_{2u}+\omega_{4u}\ge 1,\ \omega_{4l}+\omega_{1u}+\omega_{2u}+\omega_{3u}\ge 1, \\
\omega_{1u}+\omega_{2l}+\omega_{3l}+\omega_{4l}\ge 1, \omega_{2u}+\omega_{1l}+\omega_{3l}+\omega_{4l}\ge 1, \\
\omega_{3u}+\omega_{1l}+\omega_{2l}+\omega_{4l}\ge 1, \omega_{4u}+\omega_{1l}+\omega_{2l}+\omega_{3l}\ge 1, \\
\omega_{1u}-\omega_{1l}\ge 0, \omega_{2u}-\omega_{2l}\ge 0, \\
\omega_{3u}-\omega_{3l}\ge 0, \omega_{4u}-\omega_{4l}\ge 0, \\
\omega_{il}\ge 0, \omega_{iu}\ge 0, i, j=1,2,3,4
\end{cases}$$

Step 2: By utilizing the *'MatlabOptimizationToolbox'*, the solution to model (3.45) is given below:

$$\omega_{1l}=0.4778; \omega_{1u}=0.6723; \omega_{2l}=0.0135;$$
$$\omega_{2u}=0.0157; \omega_{3l}=0.2334; \omega_{3u}=0.4063;$$
$$\omega_{4l}=0.0718; \omega_{4u}=0.1216.$$

Step 3: The priority vector of the collective R_s is constructed as

$$[0.4778,0.6723] \underset{100\%}{\geq} [0.2334,0.4063] \underset{100\%}{\geq} [0.0718,0.1216] \underset{100\%}{\geq} [0.0135,0.0157]$$

Step 4: By utilizing the comparative method of interval fuzzy numbers (Horn and Johnson, 1985), we get the optimal ranking of the decision alternatives as follows:

$$x_2 \underset{100\%}{\succ} x_1 \underset{100\%}{\succ} x_3 \underset{100\%}{\succ} x_4.$$

3.6 Conclusions

Consistency properties and priority methods have represented some of the elementary and important issues in the fields of decision makers' judgments. In this chapter, we first constructed the transformation relationship between IFNRPRs and the INFPRs. Then, we developed the properties of the additively consistent and the multiplicatively consistent INFPRs, while showing the relationship among these properties.

Basing on the conditions of additive consistency of IFNCPRs, we developed the least squares optimal model (3.31) to obtain the priority vector for the collective IFNCPRs. This model embodies two kinds of significance: If $d_{ij} = m$, that is, the entries of the IFNCPRs are complete, then we obtain a collectively optimal model with complete information; and if $d_{ij} = 1$, that is, there is only one decision maker who presents his judgment information, then we obtain an individual optimal model.

Based on the conditions of multiplicative consistency of IFNCPRs, we developed the logarithmic least squares optimal model (Gong and Cui, 2008; Gong and Yao, 2009) to obtain the priority vector for the collective IFNCPRs. This model embodies three kinds of significance: If the entries of the IFNCPRs are complete, then we have a collectively optimal model with complete information; if the entries of the IFNCPRs are incomplete, then we have a collectively optimal model of incomplete information; and if there is only one decision maker who provides the judgment information, then we have an individual optimal model.

Chapter 4
Complementary Preference Relations of Triangular Fuzzy Numbers

The concepts of triangular fuzzy numbers and interval fuzzy numbers are generalizations of that of crisp numbers. A triangular fuzzy number is made up of a triple of number, including the minimum membership degree, the medium membership degree, and the maximum membership degree. In any situation of subjective decision making, this triple can be respectively regarded as the lower (conservative), the medium and the upper (optimistic) estimation with regard to a judgment. The particular form of the triangular fuzzy number safely embodies the decision maker's subjective thinking.

The comparative methods of two triangular fuzzy numbers are needed in almost all decision making. Therefore, the research on operational laws of triangular fuzzy numbers is still a hot topic in recent years. Comparing with those of intervals, operational laws of triangular fuzzy numbers are more mature and easy to use. In fact, the concept of triangular fuzzy numbers is also a generalization of that of intervals. In this sense, the related theories about preference relations with triangular fuzzy numbers possess the general significance (Laarhoven and Pedrycz, 1983; Kwiesielewicz, 1998; Wan and Sheng, 2004; Gisella and Roberto, 2004; Gao, 1999; Xiao and Li, 2003; Yu and Fan, 2004; Jiang and Fan, 2002a; 2002b; Xu, 2003; 2004; Chang, 1996; Zhu, Jing et al., 1999; Gogus and Boucher, 1998; Gong and Liu, 2006a; 2006b; Chiclana, Herrera et al., 2003; Gong, Zhang et al., 2008; Zhang, Li et al., 2006).

There are also two kinds of preference relations of triangular fuzzy numbers. One is the kind of reciprocal preference relations of triangular fuzzy numbers, written TFNRPR for short; and the other complementary preference relations of triangular fuzzy numbers, denoted TFCRPR for short. In this chapter, we focus the properties of consistency and priority methods of TFCRPRs (Jiang and Fan, 2002a; 2002b; Xu, 2003; 2004; Chang, 1996; Zhu, Jing et al., 1999; Gogus and Boucher, 1998; Gong and Liu, 2006a; Chiclana, Herrera et al., 2003; Gong and Liu, 2006b; Gong, Zhang et al., 2008; Zhang, Li et al., 2006). In the first section, we introduce the definitions of TFNRPRs and TFCRPRs respectively. In Sections 4.2 and 4.3, we discuss the properties of additively consistent and multiplicatively consistent TFCRPRs, such as restricted max-max transitivity, restricted max-min transitivity,

Z. Gong et al.: Uncertain Fuzzy Preference Relations, STUDFUZZ 281, pp. 45–74.
springerlink.com © Springer-Verlag Berlin Heidelberg 2013

and weak monotonicity. Also we investigate the relationship among these properties. In Section 4.4, we construct a least squared model on the TFCRPRs of incomplete information, and show that this model can be generalized to the case of the collective judgment matrices and individual matrices of complete information.

4.1 Basic Concepts

Definition 4.1 (Chang, 1996)**.** Let F(**R**) represent the collection of all fuzzy sets, where **R** is the set of all real numbers. For **M**$\in$ F(**R**), if

(1) There exists $x_0 \in \mathbf{R}$ such that $\mu_M(x_0)=1$;

(2) For all $\alpha \in [0,1]$, $A_\alpha = \{x : \mu_M(x) \geq \alpha$ is a closed interval.

then **M**$\in$ F(**R**) is called a fuzzy number.

Definition 4.2 (Xu, 2004; Chang, 1996)**.** A fuzzy number **M** on **R** is said to be triangular, if its membership function $\mu_M(x): R \rightarrow [0,1]$ is satisfies

$$\begin{cases} \dfrac{x}{m-l} - \dfrac{l}{m-l} & x \in [l,m], \\ \dfrac{x}{m-u} - \dfrac{u}{m-u} & x \in [m,u], \\ 0 & otherwise \end{cases}$$

where $l \leq m \leq u$, l and u stand respectively for the lower and upper bound of the support of $\boldsymbol{M}$, and m the modal (medium) value. Such a triangular fuzzy number can be denoted as (l,m,u).

Consider two positive triangular fuzzy numbers $M_1 = (l_1, m_1, u_1)$ and $M_2 = (l_2, m_2, u_2)$. Their arithmetic operations are as follows (Kwiesielewicz, 1998; Chang, 1996):

$$(l_1,m_1,u_1)+(l_2,m_2,u_2)=(l_1+l_2,m_1+m_2,u_1+u_2),$$
$$(l_1,m_1,u_1)\cdot(l_2,m_2,u_2)=(l_1l_2,m_1m_2,u_1u_2),$$
$$(\lambda,\lambda,\lambda)\cdot(l_1,m_1,u_1)=(\lambda l_1,\lambda m_1,\lambda u_1), \lambda>0,\ \lambda \in \boldsymbol{R} \text{ (real number)},$$
$$(l_1,m_1,u_1)^{-1}=(1/u_1,1/m_1,1/l_1),$$

$$(l_1, m_1, u_1) / (l_2, m_2, u_2) = (l_1 / u_2, m_1 / m_2, u_1 / l_2),$$
$$\ln M_1 = (\ln l_1, \ln m_1, \ln u_1); \ \exp M_1 = (\exp l_1, \exp m_1, \exp u_1).$$

And any $a \in R$ can be denoted as $\tilde{a} = (a, a, a) = (a_l, a_m, a_n)$.

Definition 4.3 (Laarhoven and Pedrycz, 1983; Kwiesielewicz, 1998; Gogus and Boucher, 1998). Let $\mathbf{A} = (\tilde{a}_{ij})_{n \times n}$ be a preference relation, where $\tilde{a}_{ij} = (a_{ijl}, a_{ijm}, a_{iju})$, $i, j \in$ N, then $\mathbf{A}$ is called a reciprocal preference relation of triangular fuzzy numbers (TFNRPR), if

$$\tilde{a}_{ii} = \tilde{1} \tag{4.1}$$

$$a_{ijl} a_{jiu} = a_{ijm} a_{jim} = a_{iju} a_{jil} = 1 \tag{4.2}$$

Definition 4.4 (Jiang and Fan, 2002a; 2002b; Xu, 2003; 2004). Let $\mathbf{R} = (\tilde{r}_{ij})_{n \times n}$ be a preference relation, where $\tilde{r}_{ij} = (r_{ijl}, r_{ijm}, r_{iju}), i, j \in$ N, then $\mathbf{R}$ is called a complementary preference relation of triangular fuzzy numbers (TFNCPR), if

$$\tilde{r}_{ii} = 0.\tilde{5} \tag{4.3}$$

$$r_{ijl} + r_{jiu} = r_{ijm} + r_{jim} = r_{iju} + r_{jil} = 1 \tag{4.4}$$

4.2 Properties and Priority of Additively Consistent Complementary Preference Relations of Triangular Fuzzy Numbers

4.2.1 A Comparative Method of Two Triangular Fuzzy Numbers

Definition 4.5. (Chang, 1996; Gogus and Boucher, 1998). Let $\tilde{M} = (M_l, M_m, M_u)$ and $\tilde{N} = (N_l, N_m, N_u)$ be two triangular fuzzy numbers. The degree of possibility for $\tilde{M} \geq \tilde{N}$ is defined as

$$v(\tilde{M} \geq \tilde{N}) = \sup_{x \geq y} min(\mu_M(x), \mu_N(y)) \tag{4.5}$$

From Definition 4.5, it follows that when a pair (x, y) exists such that $x \geq y$ and $\mu_M(x) = \mu_M(y) = 1$, we have $v(M_1 \geq M_2) = 1$. Therefore, we have (Gogus and Boucher, 1998):

$$v(\tilde{M} \geq \tilde{N}) = 1 \Leftrightarrow M_m \geq N_m \tag{4.6}$$

and

$$v(\tilde{N} \geq \tilde{M}) = hgt(M \cap N) = \frac{M_l - N_u}{(N_m - N_u) - (M_m - M_l)} \tag{4.7}$$

where the height $M \cap N$ is the membership value of the intersection point of the two numbers.

Lemma 4.1 (Gogus and Boucher, 1998; Buckley, 1985). Buckley defines a subjective θ such that

$$\tilde{M} \geq \tilde{N} \Leftrightarrow v(\tilde{M} \geq \tilde{N}) = 1, v(\tilde{N} \geq \tilde{M}) < \theta \tag{4.8}$$

$$\tilde{M} \approx \tilde{N} \Leftrightarrow min(v(\tilde{M} \geq \tilde{N}), v(\tilde{N} \geq \tilde{M})) \geq \theta \tag{4.9}$$

4.2.2 Additively Consistent Complementary Preference Relations of Triangular Fuzzy Numbers

This subsection is based on (Gong, Liu, 2006), with which and references found there interested readers please consult for more details.

Theorem 4.1. A complementary preference relation $\mathbf{R} = (\tilde{r}_{ij})_{n\times n}$ of triangular fuzzy numbers and a reciprocal preference relation $\mathbf{A} = (\tilde{a}_{ij})_{n\times n}$ of triangular fuzzy numbers can be transformed into each other by the using following formulas:

$$\tilde{r}_{ij} = 0.5 + 0.2log_3^{\tilde{a}_{ij}} \tag{4.10}$$

$$\tilde{a}_{ij} = 3^{5(\tilde{r}_{ij} - 0.5)}, i, j \in N \tag{4.11}$$

Proof. Let $\mathbf{A} = (\tilde{a}_{ij})_{n\times n}$ be a TFNRPR. From Eq. (4.10), it follows that if $\tilde{a}_{ii} = \tilde{1}$, then $\tilde{r}_{ii} = \tilde{0.5}$, and if $a_{ijl}a_{jiu} = a_{ijm}a_{jim} = a_{iju}a_{jil} = 1$, then

$$r_{ijl} + r_{jiu} = 0.5 + 0.2log_3^{a_{ijl}} + 0.5 + 0.2log_3^{a_{jiu}} = 1.$$

It is also easily to obtain that $r_{ijm} + r_{jim} = r_{iju} + r_{jil} = 1$. Thus, **R** is a TFNCPR.

Using the same method, we can transform **R** into **A** by Eq. (4.11). QED.

Definition 4.6 (Gogus and Boucher, 1998). A reciprocal preference relation $\mathbf{A} = (\tilde{a}_{ij})_{n\times n}$ of triangular fuzzy numbers is completely consistent (strong transitivity), if the following equation holds true:

$$\tilde{a}_{ij}\tilde{a}_{jk} = \tilde{a}_{ik} \tag{4.12}$$

Theorem 4.2 (Gogus and Boucher, 1998). A reciprocal preference relation $\mathbf{A} = (\tilde{a}_{ij})_{n\times n}$ of triangular fuzzy numbers is completely consistent if and only if the following equations hold true:

$$a_{ijm}a_{jkm} = a_{ikm} \tag{4.13}$$

$$\sqrt{a_{iju}a_{ijl}}\sqrt{a_{jku}a_{jkl}} = \sqrt{a_{iku}a_{ikl}} \tag{4.14}$$

In the following, we will establish the concept of additively consistent TFNCPRs.

From Eq. (4.13), we have

$$0.2log_3^{a_{ijm}} + 0.2log_3^{a_{jkm}} = 0.2log_3^{a_{ikm}}$$

so

$$(0.5 + 0.2log_3^{a_{ijm}}) + (0.5 + 0.2log_3^{a_{jkm}}) = 1 + 0.2log_3^{a_{ikm}}$$

That is, the following holds true:

$$r_{ijm} + r_{jkm} = r_{ikm} + 0.5 \tag{4.15}$$

Meanwhile, for $r_{ikm} = 1 - r_{kim}$, Eq. (4.15) can be rewritten as follows:

$$r_{ijm} + r_{jkm} + r_{kim} = 1.5 \tag{4.16}$$

From Eq. (4.14), the following equation follows readily:

$$r_{iju} + r_{ijl} + r_{jku} + r_{jkl} + r_{kiu} + r_{kil} = 3 \tag{4.17}$$

Definition 4.7. A complementary preference relation $\mathbf{R} = (\tilde{r}_{ij})_{n\times n}$ of triangular fuzzy numbers satisfying Eqs. (4.16) and (4.17) is said to have the property of additive consistency.

Definition 4.8. Let $\mathbf{R}=(\tilde{r}_{ij})_{n\times n}$ be a TFNCPR, for all $i, j, k \in N, i \neq j \neq k$, if

when $0.5 \leq \lambda < 1$, $\tilde{r}_{ij} \geq \tilde{\lambda}, \tilde{r}_{jk} \geq \tilde{\lambda}$ implies $\tilde{r}_{ik} \geq \tilde{\lambda}$, and

when $0 < \lambda \leq 0.5$, $\tilde{r}_{ij} \leq \tilde{\lambda}, \tilde{r}_{jk} \leq \tilde{\lambda}$ implies $\tilde{r}_{ik} \leq \tilde{\lambda}$,

then **R** is said to have the property of general transitivity.

Theorem 4.3. An additively consistent TFNCPR also satisfies the property of general transitivity.

Proof. Let $\mathbf{R}=(\tilde{r}_{ij})_{n\times n}$ be a TFNCPR. When $0.5 \leq \lambda \leq 1$, let $\tilde{r}_{ij} \geq \tilde{\lambda}, \tilde{r}_{jk} \geq \tilde{\lambda}$, then $v(\tilde{r}_{ij} \geq \tilde{\lambda}) = 1, v(\tilde{r}_{jk} \geq \tilde{\lambda}) = 1$. Thus, we have $r_{ijm} \geq \lambda, r_{jkm} \geq \lambda$. For $r_{ijm} + r_{jkm} = r_{ikm} + 0.5$, it follows that $r_{ikm} + 0.5 \geq 2\lambda$, and that $r_{ikm} \geq 2\lambda - 0.5 \geq \lambda$. That is, $v(\tilde{r}_{ik} \geq \tilde{\lambda}) = 1$. Thus, $\tilde{r}_{ik} \geq \tilde{\lambda}$.

Similarly, when $0 \leq \lambda \leq 0.5$, if $\tilde{r}_{ij} \leq \tilde{\lambda}, \tilde{r}_{jk} \leq \tilde{\lambda}$, then $\tilde{r}_{ik} \leq \tilde{\lambda}$ follows.

QED.

Definition 4.9. Let $\mathbf{R}=(\tilde{r}_{ij})_{n\times n}$ be a TFNCPR, for all $i, j, k \in N, i \neq j \neq k$. If the following conditions hold true:

If $\tilde{r}_{ij} \geq \tilde{0.5}$, $\tilde{r}_{jk} \geq \tilde{0.5}$ hold true, then $\tilde{r}_{ik} \geq \tilde{0.5}$ is implied; and

If $\tilde{r}_{ij} \leq \tilde{0.5}$, $\tilde{r}_{jk} \leq \tilde{0.5}$ hold true, then $\tilde{r}_{ik} \leq \tilde{0.5}$ is implied,

then **R** is said to have the property of satisfactory consistency.

In light of Theorem 4.3 and Definition 4.9, the following corollary follows readily.

Corollary 4.1. An additively consistent TFNCPR also possesses the property of satisfactory consistency.

Equivalent to Definition 4.9 is the following Definition 4.9'.

Definition 4.9'. If a TFNCPR satisfies the property of satisfactory consistency, then the corresponding preference relation of the decision alternatives $X = \{x_1, x_2, \cdots, x_n\}$, is said to satisfy the property of transitivity. That is, there exists a priority chain in the set $X = \{x_1, x_2, \cdots, x_n\}$ satisfying $x_{u_1} \geq x_{u_2} \geq \cdots \geq x_{u_n}$, where x_{u_i} denotes the ith decision alternative in the priority chain, and $x_{u_i} \geq x_{u_j}$ represents that x_{u_i} is preferred (superior) to x_{u_j}. If there exists a circulation $x_{u_{i1}} \geq x_{u_{i2}} \geq \cdots \geq x_{u_{ik}} \geq x_{u_{i1}}$, then the corresponding preference relation on the set $X = \{x_1, x_2, \cdots, x_n\}$, of decision alternatives is of not satisfying the property of transitivity, and the TFNCPR is inconsistent.

According to this definition, it can be sees that satisfactory consistency is the minimal logical requirement and a fundamental principle of fuzzy preference relations, which reflects the thinking characteristic of man (Tanino,1984; Herrera et al., 2004; Jiang and Fan, 2008). Therefore, it is very important to set up an approach that can judge whether a TFNCPR has the property of satisfactory consistency. In the following, we study the concept of preference matrices.

Definition 4.10. Let $\mathbf{R} = (\tilde{r}_{ij})_{n\times n}$ be a TFNCPR. Then $\mathbf{P} = (p_{ij})_{n\times n}$ is referred to as the preference matrix of **R**, where

$$p_{ij} = \begin{cases} 1 & \tilde{r}_{ij} > \tilde{0.5}, \\ 0 & \text{otherwise} \end{cases}$$

Theorem 4.4. Let $\mathbf{P} = (p_{ij})_{n\times n}$ be the preference matrix of $\mathbf{R} = (\tilde{r}_{ij})_{n\times n}$, $\mathbf{P}^i$ the ith sub-matrix of $\mathbf{P} = (p_{ij})_{n\times n}$, $i(i = 0, 1, \cdots, n-1, n)$. (That is, by deleting one 0 row vector and a corresponding column vector, a sub-matrix $\mathbf{P}^1$ of $\mathbf{P}^0 = \mathbf{P}$ is obtained;· · ·; by deleting one 0 row vector and a corresponding column vector of $\mathbf{P}^i$, a sub-matrix $\mathbf{P}^{i+1}$ of $\mathbf{P}^i$ is obtained;· · ·; by deleting one 0 row vector and a corresponding column vector of $\mathbf{P}^{n-1}$, a sub-matrix $\mathbf{P}^{n-2}$ of $\mathbf{P}^{n-1}$ is obtained, and $\mathbf{P}^n = (0)$.) Then for any $i(i = 0, 1, \cdots, n-1, n)$, $\mathbf{R} = (r_{ij})_{n\times n}$ is satisfactorily consistent if and only if there is a 0 row vector in $\mathbf{P}^i$.

Proof. Necessity. If $\mathbf{R} = (\tilde{r}_{ij})_{n\times n}$ has the property of satisfactory consistency, suppose that we have a set $X = \{x_1, x_2, \cdots, x_n\}$ of decision alternatives, which is associated with a priority chain $x_{u_1} \geq x_{u_2} \geq \cdots \geq x_{u_n}$, where x_{u_i} denotes the ith decision alternative in the priority chain. For x_{u_n} being the most inferior alternative, we have that $\tilde{r}_{u_n j} \leq \tilde{0.5}$, $j = 1, 2, \cdots, n$, and consequently $p_{u_n j} = 0, j = 1, \cdots, n$. That is, the entries of the u_n-th row are all 0. By deleting the u_n-th row and the u_n-th column of **P**, we obtain a sub-matrix $\mathbf{P}^1$. At this time, the priority relations of the rest of the decision alternatives have not changed, thus

$x_{u_{n-1}}$ is the most inferior alternative among the remaining alternatives. Obviously, in $\mathbf{P}^1$, the entries of the row represented by $x_{u_{n-1}}$ are all 0. By deleting the u_n th row and the u_n th column, and the u_{n-1} th row and the u_{n-1} th column of **P**, we produce $\mathbf{P}^2$. By continuing this procedure, eventually we have the $n-1$ th sub-matrix

$$\mathbf{P}^{n-1} = \begin{pmatrix} 0 & 1 \\ 0 & 0 \end{pmatrix} \text{or} \begin{pmatrix} 0 & 0 \\ 1 & 0 \end{pmatrix} \text{or} \begin{pmatrix} 0 & 0 \\ 0 & 0 \end{pmatrix}$$

where the 0 row vector is represented by x_{u_2}. By deleting the 0 row vector and the corresponding column, we produce $P^n = (0)$. So the most superior alternative x_{u_1} is selected.

Sufficiency. Let the entries of the u_n th row vector in **P** be 0. Then it is obviously that x_{u_n} is the most inferior decision alternative. Now by deleting the u_n th row and the u_n th column from **P**, we produce a sub-matrix $\mathbf{P}^1$. Let the 0 row vector be represented $x_{u_{n-1}}$. Then $x_{u_{n-1}}$ is superior to x_{u_n}. Continuing this procedure, at the end we obtain the most superior decision alternative x_{u_1}. Thus we create a priority chain $x_{u_1} \geq x_{u_2} \geq \cdots \geq x_{u_n}$, where x_{n_i} denotes the ith superior decision alternative in the set $X = \{x_1, x_2, \cdots, x_n\}$ of decision alternatives. That implies that **R** has the property of satisfactory consistency.

As a matter of fact, according to Theorem 4.4, we can obtain a priority algorithm for each satisfactorily consistent TFNCPR as follows.

Step 1. Construct the preference matrix.

Step 2. Let $i = 0$.

Step 3. Search for a 0 row vector in the sub-matrix $\mathbf{P}^i$. If such a 0 row exists, then the decision alternative represented by this row is denoted $x_{u_{n-i}}$. Then go to step 4. Otherwise go to Step 5.

Step 4. Delete the 0 row in $\mathbf{P}^i$ (if there are more than 1 such rows then randomly select one 0 row) and the corresponding column, and set $i = i + 1$. If $i = n$, then the decision alternative represented by this row is denoted x_{u_1} (That is, **R** has satisfactory consistency). End. otherwise, go to step 3.

Step 5. Conclude that **R** is inconsistent. End

Definition 4.11. Let $\mathbf{R} = (\tilde{r}_{ij})_{n\times n}$ be a TFNCPR, for all $i, j, k \in N, i \neq j \neq k$. Then $\mathbf{R}$ satisfies the property of restricted max-max transitivity, if

$$\tilde{r}_{ij} \geq \tilde{0.5},\ \tilde{r}_{jk} \geq \tilde{0.5} \Rightarrow \tilde{r}_{ik} \geq max\{\tilde{r}_{ij}, \tilde{r}_{jk}\}.$$

Definition 4.12. Let $\mathbf{R} = (\tilde{r}_{ij})_{n\times n}$ be a TFNCPR, for all $i, j, k \in N, i \neq j \neq k$. Then $\mathbf{R}$ is said to satisfy the property of restricted max-min transitivity, if

$$\tilde{r}_{ij} \geq \tilde{0.5},\ \tilde{r}_{jk} \geq \tilde{0.5} \Rightarrow \tilde{r}_{ik} \geq \min\{\tilde{r}_{ij}, \tilde{r}_{jk}\}.$$

Theorem 4.5. If a TFNCPR $\mathbf{R} = (\tilde{r}_{ij})_{n\times n}$ satisfies additive consistency, then it verified restricted max-max transitivity.

Proof. From the assumption that $\tilde{r}_{ij} \geq \tilde{0.5}$, $\tilde{r}_{jk} \geq \tilde{0.5}$, it follows that

$$v(\tilde{r}_{ij} \geq \tilde{0.5}) = 1,\ v(\tilde{r}_{jk} \geq \tilde{0.5}) = 1.$$

So we have $r_{ijm} \geq 0.5, r_{jkm} \geq 0.5$. From Eq.(4.15), it follows that $r_{ijm} + r_{jkm} = r_{ikm} + 0.5$ so that we have $r_{ikm} \geq r_{ijm}$, $r_{ikm} \geq r_{jkm}$. That is, $v(\tilde{r}_{ik} \geq \tilde{r}_{ij}) = 1, v(\tilde{r}_{ik} \geq \tilde{r}_{jk}) = 1$. Thus $\tilde{r}_{ik} \geq \tilde{r}_{ij}$, $\tilde{r}_{ik} \geq \tilde{r}_{jk}$. QED.

Corollary 4.2. Each additively consistent TFNCPR satisfies the property of restricted max-min transitivity.

Definition 4.13. Let $\mathbf{R} = (\tilde{r}_{ij})_{n\times n}$ be a TFNCPR, for all $i, j, p, l, s, n \in N$. Then $\mathbf{R}$ is said to satisfy the property of restricted weak monotonicity if $\tilde{r}_{ij} > \tilde{r}_{pl}, \tilde{r}_{jn} > \tilde{r}_{ls} \Rightarrow \tilde{r}_{in} > \tilde{r}_{ps}$.

Theorem 4.6. Each additively consistent TFNCPR satisfies the property of restricted weak monotonicity.

Proof. Let $\mathbf{R} = (\tilde{r}_{ij})_{n\times n}$ be a TFNCPR satisfying the property of additive consistency; and assume that $\tilde{r}_{ij} > \tilde{r}_{pl}$, $\tilde{r}_{jn} > \tilde{r}_{ls}$. Then

By $\tilde{r}_{ij} > \tilde{r}_{pl}$, we have $v(\tilde{r}_{ij} > \tilde{r}_{pl}) = 1$ so that $r_{ijm} \geq r_{plm}$; and from $\tilde{r}_{jn} > \tilde{r}_{ls}$, we have $v(\tilde{r}_{jn} > \tilde{r}_{ls}) = 1$ so that $r_{jnm} \geq r_{lsm}$. Therefore, we have

$$r_{ijm} + r_{jnm} \geq r_{plm} + r_{lsm}$$

and from

$$r_{ijm} + r_{jnm} = r_{inm} + 0.5,\ r_{plm} + r_{lsm} = r_{psm} + 0.5$$

we have

$$r_{inm} \geq r_{psm}$$

so that

$$v(\tilde{r}_{in} > \tilde{r}_{ps}) = 1 \tag{4.18}$$

In the following, it suffices to show

$$v(\tilde{r}_{in} > \tilde{r}_{ps}) < \theta \tag{4.19}$$

To this end, supposed that $v(\tilde{r}_{in} > \tilde{r}_{ps}) < \theta$ holds true. From Eq. (4.8), we have $\tilde{r}_{in} > \tilde{r}_{ps}$. Otherwise, $v(\tilde{r}_{in} > \tilde{r}_{ps}) \geq \theta$. Therefore, we have

$$\tilde{r}_{in} \approx \tilde{r}_{ps} \tag{4.20}$$

From Eq (4.9), we have $\tilde{r}_{in} \geq \tilde{r}_{ps}$. By choosing a suitable θ, we can get $v(\tilde{r}_{in} > \tilde{r}_{ps}) < \theta$. So Eqs. (4.19) and (4.20) imply that $\tilde{r}_{in} > \tilde{r}_{ps}$. QED.

Definition 4.14. Let $\mathbf{R} = (\tilde{r}_{ij})_{n\times n}$ be a TFNCPR, for all $i, j, p, l, s, n \in$ N. Then **R** is said to have the property of weak monotonicity if $\tilde{r}_{ij} \geq \tilde{r}_{pl}$, $\tilde{r}_{jn} \geq \tilde{r}_{ls} \Rightarrow \tilde{r}_{in} \geq \tilde{r}_{ps}$.

According to the proof of Theorem 4.6, we can readily obtain the following corollary.

Corollary 4.3. Each additively consistent TFNCPR is weakly monotonic.

4.2.3 Aggregations of Complementary Preference Relations of Triangular Fuzzy Numbers

Theorem 4.7. The arithmetic average combination of TFNCPRs is still complementary.

Proof. Let $\mathbf{R}^{(k)} = (\tilde{r}_{ij}^{(k)})_{n\times n}$ be a TFNCPR, $\sum_{k=1}^{m} \omega_k = 1, \omega_k \geq 0$ be some pre-determined weights, $k = 1, 2, \cdots, m$. Assume that $\tilde{r}_{ij} = \sum_{k=1}^{m} \omega_k \tilde{r}_{ij}^{(k)}$. Then, we have

$\tilde{r}_{ii} = \sum_{k=1}^{m} \omega_k \tilde{r}_{ii}^{(k)} = \sum_{k=1}^{m} \omega_k \tilde{0.5} = \tilde{0.5}$, $r_{ijm} + r_{jim} = \sum_{k=1}^{m} \omega_k (r_{ijm}^{(k)} + r_{jim}^{(k)}) = \sum_{k=1}^{m} \omega_k = 1$

$r_{ijl} + r_{jiu} = \sum_{k=1}^{m} \omega_k (r_{ijl}^{(k)} + r_{jiu}^{(k)}) = \sum_{k=1}^{m} \omega_k = 1$; $r_{iju} + r_{jil} = \sum_{k=1}^{m} \omega_k (r_{iju}^{(k)} + r_{jil}^{(k)}) = \sum_{k=1}^{m} \omega_k = 1$. ED.

Theorem 4.8. The arithmetic average combination of additively consistent TFNCPRs is also additively consistent.

Proof. Let $\mathbf{R}^{(k)} = (\tilde{r}_{ij}^{(k)})_{n\times n}$ be the kth additive consistent TFNCPR, $\omega_k, k = 1, 2, \cdots, l$ the weight of importance of $\mathbf{R}^{(k)}$, $k = 1, 2, \cdots, l$, such that $\sum_{k=1}^{l} \omega_k = 1, \omega_k \geq 0$. Let $\tilde{r}_{ij} = \sum_{k=1}^{m} \omega_k \tilde{r}_{ij}^{(k)}$.

According to Theorem 4.7, it suffices for us to show that $\tilde{R}$ satisfies the property of additive consistency. Because

$$r_{ijm} + r_{jkm} + r_{kim} = \sum_{k=1}^{l} \omega_k r_{ijm}^{(k)} + \sum_{k=1}^{l} \omega_k r_{jkm}^{(k)} + \sum_{k=1}^{l} \omega_k r_{kim}^{(k)} = \sum_{k=1}^{l} (\omega_k r_{ijm}^{(k)} + \omega_k r_{jkm}^{(k)} + \omega_k r_{kim}^{(k)}) = 1.5$$

by using the same method, we have

$$r_{ijl} + r_{jkl} + r_{kil} + r_{iju} + r_{jku} + r_{kiu} = 3.$$

QED.

Let $\tilde{r}_{ij}^{(k)}$ be the (i,j) entry of the k th TFNCPR $\mathbf{R}^{(k)} = (\tilde{r}_{ij}^{(k)})_{n\times n}$. $\mathbf{P} = (\tilde{r}_{ij}^{C})_{n\times n}$ is referred to as a collective preference relation that is aggregated by the OWA operator (the ordered weighted averaging operator) (Yager, 1992,1995,2004; Filev and Yager, 1995; Fan and Jiang, 2004). In this case,

$$\tilde{r}_{ij}^{C} = \Phi_Q(\tilde{r}_{ij}^{1}, \tilde{r}_{ij}^{2}, \cdots, \tilde{r}_{ij}^{m}) = \sum_{k=1}^{m} \omega_k \tilde{t}_{ij}^{k},$$

where $\tilde{t}_{ij}^{k}$ is the k th largest value in the set $\{\tilde{r}_{ij}^{1}, \tilde{r}_{ij}^{2}, \cdots, \tilde{r}_{ij}^{m}\}$, $Q(x)$ a relative non-decreasing quantifier (for details, see (Gong, 2006; Chiclana, Herrera, et al, 2003) with the following membership function:

$$Q(x) = \begin{cases} 0 & 0 \leq x \leq a \\ \dfrac{x-a}{b-a} & a \leq x \leq b \\ 1 & b \leq x \leq 1 \end{cases}$$

where $a, b \in [0,1], \omega_k = Q(k/m) - Q(k-1/m)$.

Theorem 4.9. Let $\mathbf{R}^{(k)} = (\tilde{r}_{ij}^{(k)})_{n\times n}$ be a TFNCPR, and $\mathbf{P} = (\tilde{r}_{ij}^{C})_{n\times n}$ a collective preference relation of triangular fuzzy numbers that is aggregated by using OWA operator. Then **P** is a complementary preference if and only if a + b = 1. (If $r_{ijm}^{s} = t_{ijm}^{k}$ is the sth largest value in the set $\{r_{ijm}^{i}, i = 1, 2, \cdots, m\}$, then treat $r_{ijl}^{s} = t_{ijl}^{k}, r_{iju}^{s} = t_{iju}^{k}$ also as the sth largest value in the set $\{r_{ijm}^{i}, i = 1, 2, \cdots, l\}$ and the set $\{r_{ijm}^{i}, i = 1, 2, \cdots, m\}$, respectively).

Because it is very similar to the method proposed in (Chiclana, Herrera, et al, 2003), the details of the proof are omitted here.

4.2.4 A Numerical Example

Suppose that a decision maker provides a TFNCPR on a set $X = \{x_1, x_2, x_3, x_4\}$ of decision alternatives as follows:

$$\mathbf{R} = \begin{pmatrix} (0.5,0.5,0.5) & (0.5,0.7,0.8) & (0.3,0.6,0.8) & (0.2,0.8,0.9) \\ (0.2,0.3,0.5) & (0.5,0.5,0.5) & (0.3,0.4,0.6) & (0.6,0.6,0.7) \\ (0.2,0.4,0.7) & (0.4,0.6,0.7) & (0.5,0.5,0.5) & (0.5,0.7,0.7) \\ (0.1,0.2,0.8) & (0.3,0.4,0.4) & (0.3,0.3,0.5) & (0.5,0.5,0.5) \end{pmatrix}$$

Step 1: Construct the preference matrix **P** as follows:

$$\mathbf{P} = \begin{pmatrix} 0 & 1 & 1 & 1 \\ 0 & 0 & 0 & 1 \\ 0 & 1 & 0 & 1 \\ 0 & 0 & 0 & 0 \end{pmatrix}$$

Step 2: Let $\mathbf{P}^0 = \mathbf{P}$.

Step 3: Search for the 0 row vector in $\mathbf{P}^0$. Obviously, the entries of the fourth row are all 0. So the decision alternative x_4 is the most inferior.

Step 4: By deleting the fourth row and the fourth column from $\mathbf{P}^0$, we get $\mathbf{P}^1$ as follows:

$$\mathbf{P}^1 = \begin{pmatrix} 0 & 1 & 1 \\ 0 & 0 & 0 \\ 0 & 1 & 0 \end{pmatrix}$$

Step 5: Search for the 0 row vector in $\mathbf{P}^1$. Obviously, the entries of the second row are all 0. This row is also the second row of **P**, so the decision alternative x_2 is superior to x_4.

Step 6: By deleting the fourth row and the fourth column and the second row and the second column from **P**, we get $\mathbf{P}^2$ as follows:

$$\mathbf{P}^2 = \begin{pmatrix} 0 & 1 \\ 0 & 0 \end{pmatrix}$$

Step 7: Search for the 0 row vector in $\mathbf{P}^2$. Obviously, the entries of the second row are all 0. This row is the third row of **P**, so the decision alternative x_3 is superior to x_2.

Step 8: By deleting the fourth row and the fourth column, the second row and the second column, and the third row and the third column from **P**, we get $\mathbf{P}^3$, where $\mathbf{P}^3 = (0)$.

Step 9: In light of Theorem 4.4 and the corresponding algorithm, the TFNCPR possesses the property of satisfactory transitivity; and we obtain the most superior decision alternative x_1.

Therefore, the priority chain of the set $X = \{x_1, x_2, x_3, x_4\}$ of decision alternatives is $x_1 \succ x_3 \succ x_2 \succ x_4$.

4.3 Properties and Priority of Multiplicatively Consistent Complementary Preference Relations of Triangular Fuzzy Numbers

4.3.1 Properties of Multiplicatively Consistent Complementary Preference Relations of Triangular Fuzzy Numbers

This section is mainly based on (Gong, Liu, 2006).

Definition 4.15 (Gogus and Boucher, 1998). Let $\omega = (\omega_1, \omega_2, \cdots, \omega_n)^T$ be a priority vector of a reciprocal preference relations of triangular fuzzy numbers (TFNRPR) A $= (a_{ij})_{n \times n}$. Then A is said to be a completely consistent TFNRPR, if $a_{ij} = \frac{\omega_i}{\omega_j}, i, j \in \mathrm{N}$, where $\omega_i = (\omega_{il}, \omega_{im}, \omega_{iu})$.

Notice that the difference between Definition 4.6 and Definition 4.15 is that the former is defined from the point of view of logical structures of human judgment, while the latter from the point of view of priority vectors.

Definition 4.16. A complementary preference relation of triangular fuzzy numbers (TFNCPR) $\mathrm{R}=(r_{ij})_{n\times n}$ is said to be multiplicative consistent, if

$$r_{ik}r_{kj}r_{ji}=r_{ij}r_{jk}r_{ki},\ i,\ j,\ \ k\in N, i\neq j\neq k\,.$$

In the following, we will discuss the transformation between TFNRPRs and the TFNCPRs.

Theorem 4.10. A TFNCPR $\mathrm{R}=(r_{ij})_{n\times n}$ and a TFNRPR $\mathrm{A}=(a_{ij})_{n\times n}$ can be transformed into each other by the following formula

$$r_{ij}=\left(1+a_{ji}\right)^{-1},i,j\in \mathrm{N} \tag{4.21}$$

Proof. Let $\mathrm{A}=(a_{ij})_{n\times n}$ be a TFNRPR, and let $r_{ij}=\left(1+a_{ji}\right)^{-1}$. Then $(r_{ijl},r_{ijm},r_{iju})=\frac{1}{1+(a_{jil},a_{jim},a_{jiu})}=(\frac{1}{1+a_{jiu}},\frac{1}{1+a_{jim}},\frac{1}{1+a_{jil}})$. That is, we have

$$(r_{jil},r_{jim},r_{jiu})=(\frac{1}{1+a_{iju}},\frac{1}{1+a_{ijm}},\frac{1}{1+a_{ijl}})$$

Thus we have

$$r_{ijl}+r_{jiu}=\left(1+a_{jiu}\right)^{-1}+\left(1+a_{ijl}\right)^{-1}=\left(1+a_{jiu}\right)^{-1}+\left(1+\frac{1}{a_{jiu}}\right)^{-1}=1$$

Similarly, we can obtain

$$r_{ijm}+r_{jim}=1 \text{ and } r_{iju}+r_{jil}=1$$

It is readily seen that $r_{iju}\geq r_{ijm}\geq r_{ijl}$ holds true for all $i,j\in N$. This shows that $\mathrm{R}=(r_{ij})_{n\times n}$ is a TFNCPR.

Similarly, if $\mathrm{R}=(r_{ij})_{n\times n}$ is a TFNCPR, then we can transform $\mathrm{R}=(r_{ij})_{n\times n}$ into $\mathrm{A}=(a_{ij})_{n\times n}$ by using $a_{ij}=r_{ji}^{-1}-1$. QED.

In the following, we investigate the concept of multiplicative consistency of TFNCPR. To this end, let $V=(v_1,\cdots,v_n)^T$ be a priority vector of a multiplicatively consistent TFNRPR A $=(a_{ij})_{n\times n}$. Obviously, we have $a_{ij}=\frac{v_i}{v_j}$, where $v_i=(v_{il},v_{im},v_{iu}), i,j\in N$. Let $r_{ij}=\frac{1}{1+a_{ji}}$. Then we have

$$a_{ij}=\frac{1}{1+\frac{v_j}{v_i}}=(\frac{v_{il}}{v_{il}+v_{ju}},\frac{v_{im}}{v_{im}+v_{jm}},\frac{v_{iu}}{v_{jl}+v_{iu}})$$

Hence, if $\omega=(\omega_1,\omega_2,\cdots,\omega_n)^T$ is the priority vector of the TFNCPR R$=(r_{ij})_{n\times n}$, then the following holds true:

$$r_{ij}=(r_{ijl},r_{ijm},r_{iju})=(\frac{\omega_{il}}{\omega_{il}+\omega_{ju}},\frac{\omega_{im}}{\omega_{im}+\omega_{jm}},\frac{\omega_{iu}}{\omega_{jl}+\omega_{iu}})$$

Obviously, the equation $r_{ij}r_{jk}r_{ki}=r_{ji}r_{kj}r_{ik}$ does not hold, for all $i,j,k\in$ N. Let us consider the following equation

$$\frac{1}{r_{ij}}-1=(\frac{1}{r_{iju}}-1,\frac{1}{r_{ijm}}-1,\frac{1}{r_{ijl}}-1)=(\frac{\omega_{jl}}{\omega_{iu}},\frac{\omega_{jm}}{\omega_{im}},\frac{\omega_{ju}}{\omega_{il}})$$

So, we have

$$(\frac{1}{r_{iju}}-1)(\frac{1}{r_{jku}}-1)(\frac{1}{r_{kiu}}-1)=\frac{\omega_{jl}}{\omega_{iu}}\frac{\omega_{kl}}{\omega_{ju}}\frac{\omega_{il}}{\omega_{ku}}=\frac{\omega_{il}}{\omega_{ju}}\frac{\omega_{jl}}{\omega_{ku}}\frac{\omega_{kl}}{\omega_{iu}}=(\frac{1}{r_{jiu}}-1)(\frac{1}{r_{kju}}-1)(\frac{1}{r_{iku}}-1) \tag{4.22}$$

Similarly, we can obtain that

$$(\frac{1}{r_{ijl}}-1)(\frac{1}{r_{jkl}}-1)(\frac{1}{r_{kil}}-1)=(\frac{1}{r_{jil}}-1)(\frac{1}{r_{kjl}}-1)(\frac{1}{r_{ikl}}-1) \tag{4.23}$$

and

$$(\frac{1}{r_{ijm}}-1)(\frac{1}{r_{jkm}}-1)(\frac{1}{r_{kim}}-1)=(\frac{1}{r_{jim}}-1)(\frac{1}{r_{kjm}}-1)(\frac{1}{r_{ikm}}-1) \tag{4.24}$$

That is, we have

$$(\frac{1}{r_{ij}}-1)(\frac{1}{r_{jk}}-1)(\frac{1}{r_{ki}}-1)=(\frac{1}{r_{ji}}-1)(\frac{1}{r_{kj}}-1)(\frac{1}{r_{ik}}-1) \tag{4.25}$$

Thus we have the concept of multiplicatively consistent TFNCPR as follows.

Definition 4.17. A TFNCPR R $=(r_{ij})_{n\times n}$ is said to have the property of multiplicative consistency, if for all $i, j, k \in N$, Eq. (4.25) holds true. And, any matrix that satisfies the property of multiplicative consistency is said to be a multiplicatively consistent matrix.

As is well known, a consistent preference must satisfy the properties of transitivity, such as restricted max-max transitivity, general transitivity, weak transitivity, and others (Tanino, 1984; Herrera et al., 2004). In the following, we will generalize the properties of transitivity of FPR (Tanino,1984; Herrera et al., 2004) to the case of TFNCPRs. Meanwhile, the relationship between the property of multiplicative consistency and the properties of transitivity of TFNCPRs is constructed to show that the defined concept of consistency is reasonable.

Definition 4.18. Let $R=(r_{ij})_{n\times n}$ be a TFNCPR such that for all $i, j, k \in N, i \neq j \neq k$, the following hold true: $r_{ij} \geq (\frac{1}{2},\frac{1}{2},\frac{1}{2}), r_{jk} \geq (\frac{1}{2},\frac{1}{2},\frac{1}{2})$.

(1) If $r_{ijl} \geq r_{jkl}, r_{ijm} \geq r_{jkm}, r_{iju} \geq r_{jku}$ imply $r_{ikl} \geq r_{ijl}$ or $r_{iku} \geq r_{iju}$ or $r_{ikm} \geq r_{ijm}$; and

(2) If $r_{jkl} \geq r_{ijl}, r_{jkm} \geq r_{ijm}, r_{jku} \geq r_{iju}$ imply $r_{ikl} \geq r_{jkl}$ or $r_{iku} \geq r_{jku}$ or $r_{ikm} \geq r_{jkm}$,

then R is said to satisfy the property of restricted max-max transitivity.

Definition 4.18 implies that if the degree with which a decision alternative X_i is preferred to alternative X_j is the triangular fuzzy number r_{ij}, and the degree with which the decision alternative X_j is preferred to alternative X_k is the triangular fuzzy number r_{jk}, then the degree with which the alternative X_i is preferred to the alternative X_k is at least the lower limit of r_{ik} or the upper limit of r_{ik} or the medium value of r_{ik}.

Theorem 4.11. Each multiplicatively consistent TFNCPR R $=(r_{ij})_{n\times n}$ satisfies the property of restricted max-max transitivity.

Proof. Let $r_{ij} \geq (\frac{1}{2},\frac{1}{2},\frac{1}{2}), r_{jk} \geq (\frac{1}{2},\frac{1}{2},\frac{1}{2})$. Then, we only need to show that

$r_{jkl} \geq r_{ijl}, r_{jkm} \geq r_{ijm}, r_{jku} \geq r_{iju}$ imply either $r_{ikl} \geq r_{jkl}$ or $r_{iku} \geq r_{jku}$ or $r_{ikm} \geq r_{jkm}$.

Suppose for the purpose of producing a contradiction that there exist $i_0, j_0, k_0 \in N, i_0 \neq j_0 \neq k_0$, such that $r_{j_0k_0l} \geq r_{i_0j_0l}, r_{j_0k_0m} \geq r_{i_0j_0m}, r_{j_0k_0u} \geq r_{i_0j_0u}$ hold true. So we have $r_{i_0k_0l} < r_{j_0k_0l}, r_{i_0k_0u} < r_{j_0k_0u}, r_{i_0k_0m} < r_{j_0k_0m}$, and $1 - r_{i_0k_0l} > 1 - r_{j_0k_0l}$. That is, $r_{k_0i_0u} > r_{k_0j_0u}$. Thus we have

$$\frac{1}{r_{i_0k_0u}} - 1 > \frac{1}{r_{j_0k_0u}} - 1, \frac{1}{r_{k_0j_0u}} - 1 > \frac{1}{r_{k_0i_0u}} - 1 \tag{4.26}$$

From $r_{i_0j_0} \geq (\frac{1}{2}, \frac{1}{2}, \frac{1}{2})$, it follows $r_{j_0i_0} \leq (\frac{1}{2}, \frac{1}{2}, \frac{1}{2})$. Thus we have

$$0 < \frac{1}{r_{i_0j_0u}} - 1 \leq 1, \frac{1}{r_{j_0i_0u}} - 1 \geq 1 \tag{4.27}$$

By applying Eqs. (4.22) and (4.27), we have

$$1 \cdot (\frac{1}{r_{j_0k\ u}} - 1)(\frac{1}{r_{k_0i_0u}} - 1) \geq (\frac{1}{r_{i_0j_0u}} - 1)(\frac{1}{r_{j_0k_0u}} - 1)(\frac{1}{r_{k_0i_0u}} - 1) = (\frac{1}{r_{j_0i_0u}} - 1)(\frac{1}{r_{k_0j_0u}} - 1)(\frac{1}{r_{i_0k_0u}} - 1) \geq 1 \cdot (\frac{1}{r_{k_0j_0u}} - 1)(\frac{1}{r_{i_0k_0u}} - 1)$$

That is, we have

$$(\frac{1}{r_{j_0k_0u}} - 1)(\frac{1}{r_{k_0i_0u}} - 1) \geq (\frac{1}{r_{k_0j_0u}} - 1)(\frac{1}{r_{i_0k_0u}} - 1) \tag{4.28}$$

However, from Eq.(4.26), we have

$$(\frac{1}{r_{j_0k_0u}} - 1)(\frac{1}{r_{k_0i_0u}} - 1) < (\frac{1}{r_{k_0j_0u}} - 1)(\frac{1}{r_{i_0k_0u}} - 1) \tag{4.29}$$

which contradicts with Eq. (4.29). That is, the condition $r_{i_0k_0l} \geq r_{j_0k_0l}, r_{i_0k_0u} \geq r_{j_0k_0u}$ holds true for all $i, j, k \in N, i \neq j \neq k$. Similarly, we can prove that $r_{i_0k_0m} \geq r_{j_0k_0m}$ holds true.

Theorem 4.11 indicates that for all $i, j, k \in N, i \neq j \neq k$, the upper limit of the membership degree satisfies $r_{iku} \geq \max\{r_{iju}, r_{jku}\}$, the lower limit of the membership degree satisfies $r_{ikl} \geq \max\{r_{ijl}, r_{jkl}\}$, and the medium value of the membership degree satisfies $r_{ikm} \geq \max\{r_{ijm}, r_{jkm}\}$. QED.

Definition 4.19. Let $R=(r_{ij})_{n\times n}$ be a TFNCPR. If for all $i, j, k \in N, i \neq j \neq k$,

when $0.5 \leq \lambda \leq 1$, $r_{ij} \geq (\lambda,\lambda,\lambda), r_{jk} \geq (\lambda,\lambda,\lambda)$ imply $r_{iku} \geq \lambda$; and

when $0 < \lambda \leq 0.5$, $r_{ij} \leq (\lambda,\lambda,\lambda), r_{jk} \leq (\lambda, \lambda, \lambda)$ imply $r_{ikl} \leq \lambda$,

then R is said to satisfy the property of general consistency.

Definition 4.19 implies that if the degree with which a decision alternative X_i is preferred to another decision alternative X_j satisfies that the lower limit $r_{ij} \geq (\lambda,\lambda,\lambda)$, and the degree with which the alternative X_j is preferred to a third decision alternative X_k satisfies that the lower limit $r_{jk} \geq (\lambda,\lambda,\lambda)$, then the degree with which the decision alternative X_i is preferred to the alternative X_k at least satisfies that the upper limit $r_{iku} \geq \lambda$.

Theorem 4.12. Each multiplicatively consistent TFNCPR satisfies the property of general consistency.

Proof. Let $R=(r_{ij})_{n\times n}$ be a multiplicatively consistent TFNCPR. For all $i, j, k \in N$, $i \neq j \neq k$, when $0.5 \leq \lambda \leq 1$, if $r_{ij} \geq (\lambda,\lambda,\lambda), r_{jk} \geq (\lambda,\lambda,\lambda)$, then from Theorem 4.11 it follows that $r_{iku} \geq \lambda$. When $0 < \lambda \leq 0.5$, if $r_{ij} \leq (\lambda,\lambda,\lambda)$, $r_{jk} \leq (\lambda,\lambda,\lambda)$, it follows that $r_{ji} \geq (1-\lambda, 1-\lambda, 1-\lambda)$, and $r_{kj} \geq (1-\lambda, 1-\lambda, 1-\lambda)$. In the following, we will verify that $r_{ikl} \leq \lambda$.

Suppose for the purpose of producing a contradiction that there exist $i_0, j_0, k_0 \in N, i_0 \neq j_0 \neq k_0$, such that $r_{i_0k_0l} > \lambda$. That is, $r_{i_0k_0} > (\lambda,\lambda,\lambda)$, $r_{k_0i_0} < (1-\lambda, 1-\lambda, 1-\lambda)$. Obviously, we have

$$\frac{1}{r_{i_0j_0l}}-1 \geq \frac{1}{\lambda}-1,\ \frac{1}{r_{j_0k_0l}}-1 \geq \frac{1}{\lambda}-1, \frac{1}{r_{j_0i_0l}}-1 \leq \frac{1}{1-\lambda}-1,\ \frac{1}{r_{k_0j_0l}}-1 \leq \frac{1}{1-\lambda}-1 \quad (4.30)$$

So, from Eqs. (4.23) and (4.30), we have

$$(\frac{\lambda}{1-\lambda})(\frac{\lambda}{1-\lambda})(\frac{1}{r_{i_0k_0l}}-1) \geq (\frac{1}{r_{j_0i_0l}}-1)(\frac{1}{r_{k_0j_0l}}-1)(\frac{1}{r_{i_0k_0l}}-1)$$

$$=(\frac{1}{r_{i_0j_0l}}-1)(\frac{1}{r_{j_0k_0l}}-1)(\frac{1}{r_{k_0i_0l}}-1) \geq (\frac{1-\lambda}{\lambda})(\frac{1-\lambda}{\lambda})(\frac{1}{r_{k_0i_0l}}-1)$$

which is equivalent to

$$\frac{1}{r_{i_0k_0l}}-1\geq(\frac{1}{r_{k_0i_0l}}-1)\frac{(1-\lambda)^4}{\lambda^4}$$

From $\frac{1}{r_{k_0i_0l}}-1>\frac{1}{1-\lambda}-1$, it follows that $r_{i_0k_0l}<\lambda^3[\lambda^3+(1-\lambda)^3]^{-1}$. So, it can be readily shown that the following inequalities

$$0<\lambda\leq 0.5 \text{ and } \lambda^3[\lambda^3+(1-\lambda)^3]^{-1}\leq\lambda$$

hold true. That is, we have

$$r_{i_0k_0l}<\lambda^3[\lambda^3+(1-\lambda)^3]^{-1}\leq\lambda<r_{i_0k_0l}$$

which is a contradiction. Thus we have $r_{ikl}\leq\lambda$. QED.

Theorem 4.12 shows that when $0.5\leq\lambda\leq 1$, if $r_{ij}\geq(\lambda,\lambda,\lambda)$, $r_{jk}\geq(\lambda,\lambda,\lambda)$, then the upper value $r_{iku}\geq\lambda$. When $0<\lambda\leq 0.5$, if $r_{ij}\leq[\lambda,\lambda], r_{jk}\leq[\lambda,\lambda]$, then the lower value $r_{ikl}\leq\lambda$.

Definition 4.20. Let $R=(r_{ij})_{n\times n}$ be a TFNCPR. If for all $i,j,k\in N,\ i\neq j\neq k$,

$$r_{ij}\geq(\frac{1}{2},\frac{1}{2},\frac{1}{2}), r_{jk}\geq(\frac{1}{2},\frac{1}{2},\frac{1}{2})\Rightarrow r_{iku}\geq\frac{1}{2}$$

or

$$r_{ij}\leq(\frac{1}{2},\frac{1}{2},\frac{1}{2}), r_{jk}\leq(\frac{1}{2},\frac{1}{2},\frac{1}{2})\Rightarrow r_{ikl}\leq\frac{1}{2}$$

then R is said to satisfy the property of weak consistency.

Corollary 4.4. Each multiplicative consistent TFNCPR satisfies the property of weak consistency.

Proof. The conclusion follows immediately from Theorem 4.12, if we let $\lambda=\frac{1}{2}$.

QED.

4.3.2 Priorities of TFNCPRs with Multiplicative Consistency

Let $\omega = (\omega_1\ \omega_2\ \cdots\ \omega_n)^T$ be the priority vector of a TFNCPR $R = (r_{ij})_{n\times n}$, where $\omega_i = (\omega_{il}, \omega_{im}, \omega_{iu})$, $i \in N$. If $R = (r_{ij})_{n\times n}$ is multiplicative consistent, then for all $i, j \in N$, we have

$$r_{ij} = (r_{ijl}, r_{ijm}, r_{iju}) = (\omega_{il}(\omega_{il} + \omega_{ju})^{-1}, \omega_{im}(\omega_{im} + \omega_{jm})^{-1}, \omega_{iu}(\omega_{jl} + \omega_{iu})^{-1})$$

That is, the following holds true:

$$\begin{cases} r_{ijl}(\omega_{il} + \omega_{ju}) = \omega_{il} \\ r_{ijm}(\omega_{im} + \omega_{jm}) = \omega_{im}, i, j \in N \\ r_{iju}(\omega_{iu} + \omega_{jl}) = \omega_{iu} \end{cases} \tag{4.31}$$

In a real-life decision making situation, it is hard for the decision makers to produce consistent TFNCPRs. That means that Eqs. (4.31) do not hold true in general. So we introduce the following deviation function:

$$\begin{cases} g_{ijl} = [r_{ijl}(\omega_{il} + \omega_{ju}) - \omega_{il}]^2 \\ g_{ijm} = [r_{ijm}(\omega_{im} + \omega_{jm}) - \omega_{im}]^2 \\ g_{iju} = [r_{iju}(\omega_{iu} + \omega_{jl}) - \omega_{iu}]^2 \end{cases}$$

It is clear that the smaller the deviation function value is, the better the consistency of judgment. Therefore, we can construct the following optimization **Model 4.1**:

$$\begin{cases} min\ g_{ijl} = [r_{ijl}(\omega_{il} + \omega_{ju}) - \omega_{il})]^2 \\ min\ g_{ijm} = [r_{ijm}(\omega_{im} + \omega_{jm}) - \omega_{im}]^2 \\ min\ g_{iju} = [r_{iju}(\omega_{iu} + \omega_{jl}) - \omega_{iu}]^2 \\ \quad 0 < \omega_{il} \le \omega_{im} \le \omega_{iu} \le 1 \\ 0 < \sum_{i=1}^{n} \omega_{jl} \le 1 \le \sum_{i=1}^{n} \omega_{ju}, i, j \in N \end{cases}$$

Because there is no preference between the deviation function values g_{ijl}, g_{ijm}, and g_{iju}, for all $i, j \in N$, we construct the following nonlinear programming **Model 4.2**:

$$[r_{ijl}(\omega_{il}+\omega_{ju})-\omega_{il}]^2+[r_{ijm}(\omega_{im}+\omega_{jm})-\omega_{im}]^2+[r_{iju}(\omega_{iu}+\omega_{jl})-\omega_{iu}]^2$$

$$s.t.\quad \begin{cases} 0<\omega_{il}\le\omega_{im}\le\omega_{iu}\le 1 \\ 0<\sum_{i=1}^{n}\omega_{il}\le 1\le\sum_{i=1}^{n}\omega_{iu},\ i\in N \end{cases}$$

4.3.3 A Numerical Example

Consider a set $\{X_1, X_2, X_3\}$ of decision alternatives. Assume that the TFNCPR provided by the decision maker is as follows:

$$\begin{pmatrix} (0.5,0.5,0.5) & (0.4,0.6,0.8) & (0.3,0.6,0.8) \\ (0.2,0.4,0.6) & (0.5,0.5,0.5) & (0.2,0.5,0.6) \\ (0.2,0.4,0.7) & (0.4,0.5,0.8) & (0.5,0.5,0.5) \end{pmatrix}$$

By utilizing the nonlinear programming software 'LINGO ', we obtain the solution to Model 4.2 as follows:

$$(\omega_{1l},\omega_{1m},\omega_{1u})=(0.20.0.43,0.61),(\omega_{2l},\omega_{2m},\omega_{2u})=(0.14.0.28,0.28),$$
$$(\omega_{3l},\omega_{3m},\omega_{3u})=\ (0.17,0.29,0.50)$$

Let us use the following equation (Xu, 2004)

$$\omega_i^{(\alpha)}=\frac{1}{2}[(1-\alpha)\omega_{il}+\omega_{im}+\alpha\omega_{iu}],\ i=1,2,3$$

to compute the expected value of triangular fuzzy numbers $\omega_i, i=1,2,3$, where α is a risk factor that measures the attitude of the decision maker. If $\alpha>0.5$, we say that the decision maker is a risk taker; if $\alpha=0.5$, the deicison maker is risk neutral; and if $\alpha<0.5$, the decision maker is risk averse. We can readily establish the $\omega_i^{(\alpha)}$ values, $i=1,2,3$, as follows:

$$\omega_1^\alpha=0.32+0.20\alpha;\ \omega_2^\alpha=0.21+0.07\alpha;\ \omega_3^\alpha=0.23+0.17\alpha$$

For all $0\le\alpha\le1$, we have

$$\omega_1>\omega_3>\omega_2$$

Notice that if we use the comparative method as studied in Subsection 4.2.1, we can also produce the same ranking result.

4.4 Group Decision Making Based on *TFNCPRs*

4.4.1 Least Squares Priority Model of Incomplete TFNCPRs

Based mainly on (Gong, Zhang et al., 2008) let us suppose that the pairwise comparisons of ***n*** decision alternatives $X_1, X_2, \cdots, X_n$ are given by *m* decision makers. For simplicity, let us write $N = \{1, 2, \cdots, n\}$ and $M = \{1, 2, \cdots, m\}$. Consider the following *TFNCPR*s:

$$\boldsymbol{R}_s = \begin{pmatrix} r_{11s} & r_{12s} & \cdots & r_{1ns} \\ r_{21s} & r_{22s} & \cdots & r_{21ns} \\ \cdots & \cdots & \cdots & \cdots \\ r_{n1s} & r_{n2s} & \cdots & r_{nns} \end{pmatrix}, s \in M .$$

The entry r_{ijs} denotes the pairwise preference degree of the decision alternative r_{ijs} over X_j, as estimated by the s th decision maker, and a_{ijs} takes the form of a triangular fuzzy number, for all $i, j \in N, s \in M$.

In practical cases, many of the decision makers may provide incomplete information. Therefore, for some $i_0, j_0 \in N$, $s_0 \in M$, $r_{i_0 j_0 s_0}$ may be empty. In this case, we denote $r_{i_0 j_0 s_0} = -$. Here we use d_{ij} to indicate the fact that there are d_{ij} pairwise comparisons between the decision alternative X_i and X_j. Obviously, $0 \le d_{ij} \le m$.

Let $W = (w_1, w_2 \cdots, w_n)^T$ be a priority vector of the decision alternatives $X_1, X_2, \cdots, X_n$ derived by the completely consistent TFNRPR $R = (r_{ij})_{n \times n}$, then we have $a_{ij} = w_i / w_j$, where $w_i, i \in N$ is a positive triangular fuzzy number. According to equation (4.10), if we let $V = (v_1, v_2 \cdots, v_n)^T$ be a priority vector of the completely consistent TFNCPR $R_s = (r_{ijs})_{n \times n}$, then we have

$$r_{ijs} = 0.5 + 0.2 log_3^{v_i / v_j} \tag{4.32}$$

where $v_i, i \in N$ is a positive triangular fuzzy number.

However, in the general case, Eq. (4.32) cannot hold true for any inconsistent estimations. So we get an optimal priority $V = (v_1, v_2 \cdots, v_n)^T$ by minimizing the squared errors between r_{ijs} and $0.5 + 0.2 log_3^{v_i / v_j}$, for all $i, j \in N, s \in M$, as follows, where it is assumed that all the decision makers are of the same importance:

$$min \quad \sum_{i=1}^{n}\sum_{j=1}^{n}\sum_{s=1}^{d_{ij}}(r_{ijs} - 0.5 - 0.2log_3^{v_i/v_j})^2 \tag{4.33}$$

Let $y_{ijs} = r_{ijs} - 0.5$ and $x_i = 0.2log_3^{v_i}$. Then Eq. (4.33) is equivalent to

$$min \quad \sum_{i=1}^{n}\sum_{j=1}^{n}\sum_{s=1}^{d_{ij}}(y_{ijs} - x_i + x_j)^2 \tag{4.34}$$

The optimal solution to Eq. (4.34) is given as follows:

$$x_i \sum_{j=1, j\neq i}^{n} d_{ij} - \sum_{j=1, j\neq i}^{n} d_{ij}x_j = \sum_{j=1, j\neq i}^{n}\sum_{s=1}^{d_{ij}} y_{ijs}, i \in \mathrm{N} \tag{4.35}$$

where $y_{ijs} = (l_{ijs}, m_{ijs}, u_{ijs})$, $x_i = (l_i, m_i, u_i), i, j \in \mathrm{N}$, $s \in \mathrm{M}$ are positive triangular fuzzy number. Eq.(4.35) can be transformed into the following equations:

$$i \sum_{j=1, j\neq i}^{n} d_{ij} - \sum_{j=1, j\neq i}^{n} d_{ij}u_j = \sum_{j=1, j\neq i}^{n}\sum_{s=1}^{d_{ij}} l_{ijs}, i \in \mathrm{N} \tag{4.36}$$

$$m_i \sum_{j=1, j\neq i}^{n} d_{ij} - \sum_{j=1, j\neq i}^{n} d_{ij}m_j = \sum_{j=1, j\neq i}^{n}\sum_{s=1}^{d_{ij}} m_{ijs}, i \in \mathrm{N} \tag{4.37}$$

$$u_i \sum_{j=1, j\neq i}^{n} d_{ij} - \sum_{j=1, j\neq i}^{n} d_{ij}l_j = \sum_{j=1, j\neq i}^{n}\sum_{s=1}^{d_{ij}} u_{ijs}, i \in \mathrm{N} \tag{4.38}$$

For simplicity, let $\sum_{j=1, j\neq i}^{n}\sum_{s=1}^{d_{ij}} m_{ijs} = a_i$, $\sum_{j=1, j\neq i}^{n}\sum_{s=1}^{d_{ij}} l_{ijs} = b_i$, and $\sum_{j=1, j\neq i}^{n}\sum_{s=1}^{d_{ij}} u_{ijs} = c_i$, $i \in \mathrm{N}$. Then the matrix format of Eqs. (4.36), (4.37) and (4.38) is

$$\boldsymbol{Dm} = \boldsymbol{a} \tag{4.39}$$

$$\boldsymbol{Eh} = \boldsymbol{d} \tag{4.40}$$

where

$$\boldsymbol{E} = \begin{pmatrix} D_1 & D_2 \\ D_2 & D_1 \end{pmatrix}$$

$$\boldsymbol{D}=\begin{pmatrix} \sum_{j=1,j\neq 1}^{n} d_{1j} & -d_{12} & \cdots & -d_{1n} \\ -d_{21} & \sum_{j=1,j\neq 2}^{n} d_{2j} & \cdots & -d_{2n} \\ \cdots & \cdots & \cdots & \cdots \\ -d_{n1} & -d_{n2} & \cdots & \sum_{j=1,j\neq n}^{n} d_{nj} \end{pmatrix}$$

$$\boldsymbol{D}_1=\begin{pmatrix} \sum_{j=1,j\neq 1}^{n} d_{1j} & 0 & \cdots & 0 \\ 0 & \sum_{j=1,j\neq 2}^{n} d_{2j} & \cdots & 0 \\ \cdots & \cdots & \cdots & \cdots \\ 0 & 0 & \cdots & \sum_{j=1,j\neq n}^{n} d_{nj} \end{pmatrix}, \boldsymbol{D}_2=\begin{pmatrix} 0 & -d_{12} & \cdots & -d_{1n} \\ -d_{21} & 0 & \cdots & -d_{2n} \\ \cdots & \cdots & \cdots & \cdots \\ -d_{n1} & -d_{n2} & \cdots & 0 \end{pmatrix}$$

$$\boldsymbol{m}=(m_1,m_2,\cdots,m_n)^T, \boldsymbol{h}=(l_1,l_2,\cdots,l_n,u_1,u_2,\cdots,u_n)^T$$

and

$$\boldsymbol{a}=(a_1,a_2,\cdots,a_n)^T, \boldsymbol{d}=(b_1,b_2,\cdots,b_n,c_1,c_2,\cdots,c_n)^T$$

By the complementary property of the preference relation, we have

$$\sum_{i=1}^{n} a_i = 0, \sum_{i=1}^{n} (b_i + c_i) = 0$$

The matrix $\boldsymbol{D}' = (d'_{ij})_{n\times n}$ actually stands for a measure for the degree of the incompleteness of the TFNCPRs: For all $i, j \in N, i \neq j$, $d'_{ij} = 0$ denotes that there is no decision maker that presents his estimation value regarding the decision alternatives X_i and X_j; $|d'_{ij}| = m$ denotes that all decision makers present their estimation values between the decision alternatives X_i and X_j; $|d'_{ij}| < m$ denotes that $|d'_{ij}|$ decision makers present their estimation values between the decision alternative X_i and X_j. As detailed in Chapter 2, only when there exist direct or indirect judgments between the decision alternatives X_i and X_j, we can rank the decision alternatives. Hence, we have the following conclusion: Let

R_s, $s \in M$, be the TFNCPRs, Δ the indicator matrix of ***D***, ***G*** be the directed graph of Δ, then ***G*** is strongly connected $\Longleftrightarrow$ Δ is irreducible $\Rightarrow$ all the decision alternatives can be ranked by utilizing some approach.

In the following, we will first discuss the existence condition of Eqs.(4.39) and (4.40), then derive the general solution. Due to similarity, we only need to focus on Eq.(4.39). Firstly, let us consider the properties of the matrix ***D***:

(1) ***D*** is a symmetric matrix;
(2) ***D*** is diagonally dominant (Horn and Johnson, 1985); and
(3) ***D*** is a semi-positive matrix (Geršgorins' discs Theorem (Yang, 1989).

Theorem 4.13. If the indicator matrix of ***D*** is irreducible, then $\mathrm{R}(\boldsymbol{D}) = n-1$.

Proof. For the reason that the sum of the entries of each row vector in ***D*** is 0, we have $\mathrm{R}(\boldsymbol{D}) \leq n-1$. By deleting the *n*th row and the *n*th column of ***D***, we get the following sub-matrix

$$\boldsymbol{D}_1 = \begin{pmatrix} \sum\limits_{j=1,j\neq 1}^{n} d_{1j} & -d_{12} & \cdots & -d_{1,n-1} \\ -d_{21} & \sum\limits_{j=1,j\neq 2}^{n} d_{2j} & \cdots & -d_{2,n-1} \\ \cdots & \cdots & \cdots & \cdots \\ -d_{n-1,1} & -d_{n-1,2} & \cdots & \sum\limits_{j=1,j\neq n}^{n} d_{n-1,j} \end{pmatrix}$$

Because the indicator matrix of ***D*** is assumed to be irreducible, the directed graph corresponding to ***D*** is strongly connected. This end denotes that there must exist $-d_{in} \neq 0, i = 1, 2, \cdots, n-1$. Otherwise, if $d_{in} = 0$, for all $i = 1, 2, \cdots, n-1$, there then is no decision alternative to be used to compare with the decision alternative X_n. That would contradict with the fact that ***D*** is strongly connected. Without loss of generality, we let $d_{1n} \neq 0$. Then we have $\sum\limits_{j=1,j\neq 1}^{n} d_{1j} > d_{12} + d_{13} + \cdots + d_{1,n-1}$. Moreover, the directed graph corresponding to $\boldsymbol{D}_1$ is also strongly connected, which implies that $\boldsymbol{D}_1$ is irreducible. According to (Yang, 1989), a diagonally dominant matrix is reversible. Thus we conclude that $\boldsymbol{D}_1$ is reversible, and $\mathrm{R}(\boldsymbol{D}) = n-1$. QED.

Corollary 4.5. If the indicator matrix Δ of ***D*** is irreducible, then equation $\boldsymbol{Dm} = \boldsymbol{a}$ has a solution.

Proof. According to Theorem 4.13, R(***D***)$=n-1$. For $\sum_{i=1}^{n} a_i = 0$, and the fact that the sum of the entries of each row vector of ***D*** is 0, we have R(***D,a***) = R(***D***)$=n-1$. This means that the equation ***Dm*** = ***a*** must have at least one solution, of which the general solution is given as follows (Yang, 1989; Horn and Johnson, 1985):

$$m = D^{+}a + (I - D^{+}D)y$$

where D^{+} stands for the generalized pseudo-inverse of ***D***, I the identity matrix of order $2n$, and ***y*** an arbitrary column vector. If we let ***y*** = 0, then we produce a particular solution to the equation ***Dm*** = ***a***, where $m = D^{+}a$.

According to Theorem 4.13, if the indicator matrix Δ of ***D*** is irreducible, then R(***D***) $=n-1$. For the reason that ***E*** is semi-positive definite, according to (Yang, 1989), there exists an orthogonal matrix ***P*** such that $D^{+}=P\wedge^{+}P^{T}$, where $\wedge$ is a diagonal matrix with $\lambda_1,\cdots,\lambda_{n-1},0$, the eigenvalues of ***E***, along diagonal, and

$$I\text{-}D^{+}D=\begin{pmatrix} 1/n & 1/n & \cdots & 1/n \\ 1/n & 1/n & \cdots & 1/n \\ \cdots & \cdots & \cdots & \cdots \\ 1/n & 1/n & \cdots & 1/n \end{pmatrix}=T$$

Thus, if the indicator matrix Δ of ***D*** is irreducible, the solution to ***Dm*** = ***a*** is

$$m\text{=}Da\text{+}Ty\text{=}\ D^{+}a + \left(\frac{1}{n}\sum_{i=1}^{n} y_i, \frac{1}{n}\sum_{i=1}^{n} y_i, \cdots, \frac{1}{n}\sum_{i=1}^{n} y_i\right)^T$$

where $y=(y_1, y_2,\cdots, y_n)^T$ is an arbitrary real vector. If we let $\frac{1}{n}\sum_{i=1}^{n} y_i = p_1$, then p_1 is also an arbitrary real number. Let $P_1=(p_1, p_1,\cdots, p_1)^T$. Then the solution to the equation ***Dm*** = ***a*** can be denoted in a more general form:

$$m\text{=}D^{+}\text{a}+P^{1} \tag{4.41}$$

Similarly, if the indicator matrix of ***E*** is irreducible, then R(***E,d***)=R(***E***)= $2n-1$. The equation ***Eh*** = ***d*** has at least one solution; and its general solution is given as follows:

$$h = E^{+}d + (I\text{-}E^{+}E)Z \tag{4.42}$$

where $\boldsymbol{E}^{+}$ is the generalized pseudo-inverse of $\boldsymbol{E}$, and z an arbitrary column vector. If we let $\boldsymbol{y}$ = 0, then we derive the following particular solution to $\boldsymbol{Eh} = \boldsymbol{d}$,

$$\boldsymbol{h} = \boldsymbol{E}^{+}\boldsymbol{d} + \boldsymbol{P}_2 \tag{4.43}$$

where $\boldsymbol{P}_2 = (p_2, p_2, \cdots, p_2)^T$ is an arbitrary real vector. QED.

It deserve to point out that even when the indicator matrix of $\boldsymbol{D}$ is irreducible, the indicator matrix of $\boldsymbol{E}$ may still be reducible and Eq. (4.40) still may have a solution. For example, let us consider the following matrices:

$$\boldsymbol{D} = \begin{pmatrix} \sum_{j=1, j\neq 1}^{n} d_{1j} & -d_{12} & \cdots & -d_{1n} \\ -d_{21} & d_{21} & \cdots & 0 \\ \cdots & \cdots & \cdots & \cdots \\ -d_{n1} & 0 & \cdots & d_{n1} \end{pmatrix}, \boldsymbol{E} = \begin{pmatrix} D_1 & D_2 \\ D_2 & D_1 \end{pmatrix}$$

$$\boldsymbol{D}_1 = \begin{pmatrix} \sum_{j=1, j\neq 1}^{n} d_{1j} & 0 & \cdots & 0 \\ 0 & d_{21} & \cdots & 0 \\ \cdots & \cdots & \cdots & \cdots \\ 0 & 0 & \cdots & d_{n1} \end{pmatrix}, \boldsymbol{D}_2 = \begin{pmatrix} 0 & -d_{12} & \cdots & -d_{1n} \\ -d_{21} & 0 & \cdots & 0 \\ \cdots & \cdots & \cdots & \cdots \\ -d_{n1} & 0 & \cdots & 0 \end{pmatrix}$$

where the diagonal, the first row, and first column of $\boldsymbol{D}$ are not all 0, and $d_{1j} = d_{j1} \neq 0, \quad j = 2, \cdots, n$. We can prove that $\boldsymbol{D}$ is irreducible, but $\boldsymbol{E}$ is reducible. And we can also prove that R($\boldsymbol{E}$,$\boldsymbol{d}$)= R($\boldsymbol{E}$)$= 2n-2$, which denotes that $\boldsymbol{Eh} = \boldsymbol{d}$ must have at least one solution.

Corollary 4.6. Suppose that the indicator matrix of $\boldsymbol{D}$ is irreducible. If the indicator matrix of $\boldsymbol{E}$ is also irreducible, then Eq. (4.34) must have at least one solution, and its general solution is given in Eqs. (4.41) and (4.43). If the indicator matrix of $\boldsymbol{E}$ is reducible, and R($\boldsymbol{E}$,$\boldsymbol{d}$) = R($\boldsymbol{E}$), then Eq. (4.35) must have at least one solution, and its general solution is given in Eqs. (4.41) and (4.42).

For $x_i = 0.2 log_3^{v_i}$, we have $v_i = 243^{x_i}, i \in$ N. Hence the general solution to Eq. (4.33) is $v_i = (243^{l_i}, 243^{m_i}, 243^{u_i})$.

4.4.2 A Numerical Example

Along with the advent of knowledge economic epic and the development trend of global economy, there have appeared more and more intense competitions among enterprises. The ability to produce technological innovation is a leading factor of measuring competitive ability of a country. Petroleum is the economic lifeline of the country, and the ability for technological innovation of petroleum enterprises has a direct bearing on the future of the nation. Setting up a scientific evaluation system for the capability of technological innovative of petroleum enterprises is very important. The decision makers usually prefer the method of qualitative evaluation to quantity method in terms of measuring the ability for technological innovation of petroleum enterprises, because these enterprises involve many aspects and departments (Zhang, Li et al., 2006a, 2006b).

Suppose that there are three decision makers d_1, d_2, and d_3 that evaluate the innovation ability of four petroleum enterprises X_1, X_2, X_3, and X_4. Their incomplete TFNCPRs $\{\tilde{R}_1, \tilde{R}_2, \tilde{R}_3\}$ are presented as follows:

$$\boldsymbol{R}_1 = \begin{pmatrix} (0.5,0.5,0.5) & (0.5,0.7,0.8) & (0.3,0.4,0.5) & (0.2,0.3,0.4) \\ (0.2,0.3,0.5) & (0.5,0.5,0.5) & 0 & 0 \\ (0.5,0.6,0.7) & 0 & (0.5,0.5,0.5) & 0 \\ (0.6,0.7,0.8) & 0 & 0 & (0.5,0.5,0.5) \end{pmatrix}$$

$$\boldsymbol{R}_2 = \begin{pmatrix} (0.5,0.5,0.5) & (0.5,0.6,0.8) & 0 & 0 \\ (0.2,0.4,0.5) & (0.5,0.5,0.5) & 0 & (0.6,0.6,0.7) \\ 0 & 0 & (0.5,0.5,0.5) & (0.5,0.7,0.7) \\ 0 & (0.3,0.4,0.4) & (0.3,0.3,0.5) & (0.5,0.5,0.5) \end{pmatrix}$$

and

$$\boldsymbol{R}_3 = \begin{pmatrix} (0.5,0.5,0.5) & (0.5,0.6,0.7) & 0 & 0 \\ (0.3,0.4,0.5) & (0.5,0.5,0.5) & 0 & (0.5,0.6,0.7) \\ 0 & 0 & (0.5,0.5,0.5) & 0 \\ 0 & (0.3,0.4,0.5) & 0 & (0.5,0.5,0.5) \end{pmatrix}$$

Step 1: Construct the matrices $\boldsymbol{D}$ and $\boldsymbol{E}$ as follows:

$$\boldsymbol{D}=\begin{pmatrix} 5 & -3 & -1 & -1 \\ -3 & 5 & 0 & -2 \\ -1 & 0 & 2 & -1 \\ -1 & -2 & -1 & 4 \end{pmatrix},\ \boldsymbol{E}=\begin{pmatrix} 5 & 0 & 0 & 0 & 0 & -3 & -1 & -1 \\ 0 & 5 & 0 & 0 & -3 & 0 & 0 & -2 \\ 0 & 0 & 2 & 0 & -1 & 0 & 0 & -1 \\ 0 & 0 & 0 & 4 & -1 & -2 & -1 & 0 \\ 0 & -3 & -1 & -1 & 5 & 0 & 0 & 0 \\ -3 & 0 & 0 & -2 & 0 & 5 & 0 & 0 \\ -1 & 0 & 0 & -1 & 0 & 0 & 2 & 0 \\ -1 & -2 & -1 & 0 & 0 & 0 & 0 & 4 \end{pmatrix}$$

Step 2: According to Eqs.(4.39) and (4.40), we establish the equations $\boldsymbol{Dm}=\boldsymbol{a}$ and $\boldsymbol{Eh}=\boldsymbol{d}$ so that the following hold:

$$\boldsymbol{a}=(0.1,-0.2,0.3,-0.2)^T,\ \boldsymbol{d}=(-0.5,-0.7,0,-0.5,0.7,0.4,0.4,0.2)^T$$

Step3: Test whether or not the matrices $\boldsymbol{D}$ and $\boldsymbol{E}$ are irreducible (or whether or not the corresponding directed graphs of the matrices $\boldsymbol{D}$ and $\boldsymbol{E}$ are stronger connected). In this example, the indicators of the matrices $\boldsymbol{D}$ and $\boldsymbol{E}$ are all irreducible.

Step 4: The generalized pseudo-inverses of $\boldsymbol{D}$ and $\boldsymbol{E}$ are found to be as follows:

$$\boldsymbol{D}^{+}=\begin{pmatrix} 0.1273 & 0.0069 & -0.0856 & -0.0486 \\ 0.0069 & 0.1458 & -0.1319 & -0.0208 \\ -0.0856 & -0.1319 & 0.2940 & -0.0764 \\ -0.0486 & -0.0208 & -0.0764 & 0.1458 \end{pmatrix}$$

$$\boldsymbol{E}^{+}=\begin{pmatrix} 0.2622 & -0.1362 & -0.1678 & 0.0272 & -0.1349 & 0.1432 & 0.0822 & -0.0758 \\ -0.1362 & 0.3009 & 0.0590 & -0.1207 & 0.1432 & -0.1550 & -0.1910 & 0.0999 \\ -0.1678 & 0.0590 & 0.5220 & -0.1632 & 0.0822 & -0.1910 & -0.2280 & 0.0868 \\ 0.0272 & -0.1207 & -0.1632 & 0.2714 & -0.0758 & 0.0999 & 0.0868 & -0.1256 \\ -0.1349 & 0.1432 & 0.0822 & -0.0758 & 0.2622 & -0.1362 & -0.1678 & 0.0272 \\ 0.1432 & -0.1550 & -0.1910 & 0.0999 & -0.1362 & 0.3009 & 0.0590 & -0.1207 \\ 0.0822 & -0.1910 & -0.2280 & 0.0868 & -0.1678 & 0.0590 & 0.5220 & -0.1632 \\ -0.0758 & 0.0999 & 0.0868 & -0.1256 & 0.0272 & -0.1207 & -0.1632 & 0.2714 \end{pmatrix}$$

Step 5: By using Eqs.(4.40) and (4.42), and letting $\boldsymbol{P}_1=\boldsymbol{P}_2=0$, we obtain the solutions to the equations $\boldsymbol{Dm}=\boldsymbol{a}$ and $\boldsymbol{Eh}=\boldsymbol{d}$ as follows:

$$\boldsymbol{D}^{+}\boldsymbol{a}=(-0.0046,-0.0639,0.1213,-0.0528)^{T}$$

$$\boldsymbol{E}^{+}\boldsymbol{d}=(-0.0688,-0.1004,0.0315,-0.0683,0.0725,0.0114,0.1315,-0.0095)^{T}$$

Therefore, the priority vector of the group decision making is

$$\begin{aligned}(v_1,v_2,v_3,v_4)^T = ((&0.6853,0.975,1.4892),(0.5761,0.704,1.0646),\\ &(1.1889,1.9470,2.0592),(0.6872,0.7482,0.9492))^T.\end{aligned}$$

According to the ranking method of triangular fuzzy numbers as proposed by (Xu and Da, 2003), we derive the priority chain $X_3 \succ X_1 \succ X_4 \succ X_2$ of the decision alternatives.

4.5 Conclusions

Consistency is a very important property for all the preference relations; it reflects how consistent a decision maker's judgment is. Basing on the transformation between TFNCPRs and TFNRPRs, in this chapter we studied the definitions of additive and multiplicative consistency for TFNCPRs, and established the concepts of restricted max-min transitivity, restricted max-max transitivity, and weak monotonicity. By using the comparative method of triangular fuzzy numbers, we studied the inherent relationships between the two kinds of consistent TFNCPRs and their properties. We also proposed an aggregation method for TFNCPRs based on OWA operator. At the same time, we established an algorithm that can be employed to judge whether or not a given TFNCPR possesses the property of satisfactory transitivity.

In this chapter, other than proposing a least squared model for group decision making based on incomplete TFNCPR, we also studied the existence condition for the solution of this model. It is also shown that the priority model of the collective judgment matrices with incomplete information can be extended to the case of the collective judgment matrices and the individual matrix with complete information.

Chapter 5
Two-Tuple Linguistic Preference Relations

In decision makings involving multiple attributes, terms used for linguistic evaluations, such as outstanding, good, and poor, are difficult to deal with in terms of quantitative models. The evaluations of these terms are often transformed into mathematical symbols (variables). In recent years, a great amount of attentions have been attracted to the decision making analysis of linguistic variables (Fan, and Jiang, 2003, 2004; Xu, 2004e,2005b,2005c; Chen and Fan, 2004; Hou and Wu, 2005a, 2005b; Jiang and Fan, 2003; Chen and Fan,2004; Herrera, 2001; Herrera, Herrera, and Martinez, 2000; Herrera and Martinez, 2000; 2001a; 2001b; Delgado, Herrera, and Herrera, 2002; Herrera, Martinez, and Sanchez, 2005; Wang and Fan, 2003; You, Fan, and Li, 2005;Gong, 2007; Gong and Liu, 2007c, Hou and Wu,2005b). The concept of linguistic preference relations (LPRs) has been one of active research fields. However, decision making outcomes out of LPRs usually do not match the meanings of the initial linguistic terms so that processes of approximation have to be employed. That implies that the eventual results of the decision making do not capture all available information. Therefore, a useful 2-tuple fuzzy linguistic representation model was suggested (Herrera, 2001; Herrera, Herrera, and Martinez, 2000; Herrera and Martinez, 2001a; Herrera and Martinez, 2000; Herrera and Martinez, 2001b; Delgado, Herrera, and Herrera, 2002; Herrera, Martinez, and Sanchez, 2005) to the study of decision making problems. The merit of the 2-tuple fuzzy linguistic representation model is that it allows one to compute with words without loss of information (Herrera, 2001; Herrera, Herrera, and Martinez, 2000; Herrera and Martinez, 2001a; Herrera and Martinez, 2000; Herrera and Martinez, 2001b; Delgado, Herrera, and Herrera, 2002; Herrera, Martinez, and Sanchez, 2005; Jiang and Fan, 2003; Wang and Fan, 2003; You, Fan, and Li, 2005; Gong and Liu, 2007c). Properties and priority rankings of 2-tuple linguistic preference relations are also among the key contents of this chapter. We first discuss the properties of 2-tuple LPRs, such as additive consistency, satisfactory consistency, max-max transitivity, and max-min transitivity, and then establish the relationship among these properties. In Section 5.3, by using a transformation relation between numerical values and two-tuple linguistic values, we prove that the transformed 2-tuple LPRs still preserve the properties of the original values. This result confirms the fact that the transformation method is feasible.

Z. Gong et al.: Uncertain Fuzzy Preference Relations, STUDFUZZ 281, pp. 75–91.
springerlink.com © Springer-Verlag Berlin Heidelberg 2013

5.1 The Basic Concept of 2-Tuple Linguistic Term Set

In many real-life scenarios of decision making, the decision makers may prefer qualitative assessments to quantitative forms. In such cases, sets of linguistic terms are often employed to express the opinions of the individual decision makers. Let S = $\left\{s_0, s_1, \ldots, s_g\right\}$ be a set of linguistic terms, where the label S_i represents a possible value of a linguistic variable, and the cardinality g, $2 \leq g \leq 14$, is usually an even number. Each linguistic value S_i is usually quantified into a triangular fuzzy number, that is $S_i = (a_i, b_i, c_i)$, where

$$a_i = \tfrac{i-1}{g}(1 \leq i \leq g),\ b_i = \tfrac{i}{g}\ (0 \leq i \leq g), c_i = \tfrac{i+1}{g}(0 \leq i \leq g-1),\quad a_0 = 0, c_g = 1.$$

An example of a set of seven linguistic terms with the associated semantics is given below, as shown in Figure 5.1:

$$S_0 = N = None = (0, 0, 0.17),\ S_1 = VL = Very\ Low = (0, 0.17, 0.33),$$
$$S_2 = L = Low = (0.17, 0.33, 0.5)\ S_3 = M = Medium = (0.33, 0.5, 0.67),$$
$$S_4 = H = High = (0.5, 0.67, 0.83),\ S_5 = VH = Very\ High = (0.67, 0.83, 1),$$
$$S_6 = P = Perfect = (0.83, 1, 1).$$

Another example of a set of nine linguistic terms with the associated semantics is given below, as shown in Figure 5.2:

$$S_0 = N = None = (0, 0, 0.125),\ S_1 = PL = Perfect\ Low = (0, 0.125, 0.25),$$
$$S_2 = VL = Very\ Low = (0.125, 0.25, 0.375),\ S_3 = L = Low = (0.25, 0.375, 0.5),$$
$$S_4 = M = Medium = (0.375, 0.5, 0.625),\ S_5 = H = High = (0.5, 0.625, 0.75),$$
$$S_6 = VH = Very\ High = (0.625, 0.75, 0.875),\ S_7 = PH = Perfect\ High = (0.75, 0.875, 1),$$
$$S_8 = P = Perfect = (0.875, 1, 1).$$

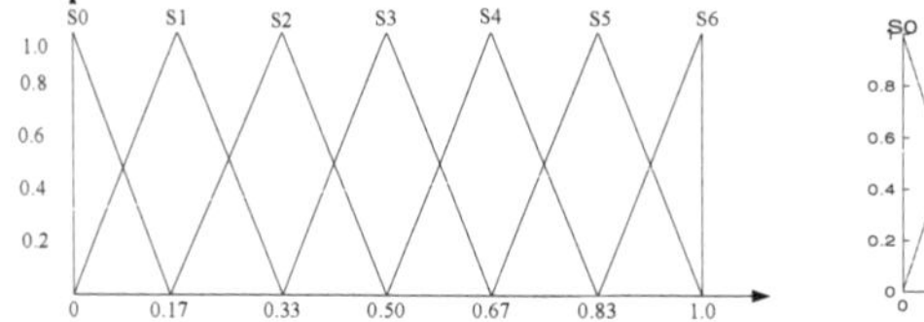

Fig. 5.1 A set of 7 terms with its semantics

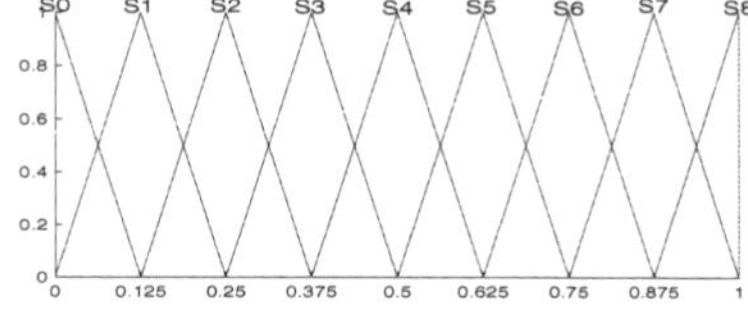

Fig. 5.2 A set of 9 terms with its semantics

Each set of linguistic terms satisfies the following properties (Herrera and Martinez, 2000):

(1) The set is ordered such that $S_i \leq S_j$, if $i < j$;
(2) There is a negation operator *Neg* satisfying $Neg(S_i) = S_j$ such that $j = g - i$;
(3) There are a maximization and a minimization operator *Max* and *Min* that satisfy

$$Max\{S_i, S_j\} = S_i; Min\{S_i, S_j\} = S_j, \text{ if } S_i \geq S_j$$

Let $S = \{s_0, \ldots, s_g\}$ be a set of linguistic terms. If a symbolic method that aggregates linguistic information produces a value $\beta \in [0, g]$ such that $\beta \notin \{0, 1, 2, \cdots, g\}$, then an approximation function can be used to express the index of the result in *S*. That implies that the eventual outcomes of the approximation do not capture all the available information. Therefore, a useful 2-tuple fuzzy linguistic representation model was suggested by Herrera et al et al. (Herrera, Martinez and Sanchez, 2005; Jiang and Fan, 2003; Wang and Fan, 2003).

Definition 5.1 (Herrera, Herrera, and Martinez, 2000; Herrera and Martinez, 2001a; Herrera and Martinez, 2000; 2001b; Herrera, 2001; Delgado, Herrera, and Herrera, 2002; Herrera, Martinez, and Sanchez, 2005). Let β be the result of an aggregation of the indexes of a set of labels that are used in a set *S* of linguistic terms, i.e., the result of a symbolic aggregation operation. In particular, $\beta \in [0, g]$, where $g+1$ is the cardinality of *S*. Let $i = round(\beta)$ and $\alpha = \beta - i$ be two values such that $i \in [0, g]$ and $\alpha \in [-0.5, 0.5)$. Then α is referred to as a symbolic translation.

Definition 5.2 (Herrera, Herrera, and Martinez, 2000; Herrera and Martinez, 2001a; Herrera and Martinez, 2000; 2001b; Herrera, 2001; Delgado, Herrera, and Herrera, 2002; Herrera, Martinez, and Sanchez, 2005). Let $S = \{s_0, \ldots, s_g\}$ be a set of linguistic terms and $\beta \in [0, g]$ a cardinal value, the result of a symbolic aggregation operation. Then the 2-tuple that expresses the information equivalent to β is obtained with the following function:

$$\Delta : [0, g] \rightarrow S \times [-0.5, 0.5)$$
$$\Delta(\beta) = (s_i, \alpha), \ where \begin{cases} s_i, & i = round(\beta); \\ \alpha = \beta - i, & \alpha \in [-0.5, 0.5). \end{cases} \tag{5.1}$$

where round (.) is the usual rounding operation, and s_i has the closest index label to β and α is the value of symbolic translation.

Definition 5.3 (Herrera, Herrera, and Martinez, 2000; Herrera and Martinez, 2001a; Herrera and Martinez, 2000; 2001b; Herrera, 2001; Delgado, Herrera, and Herrera, 2002; Herrera, Martinez, and Sanchez, 2005). Let $S = \{s_0, \ldots, s_g\}$ be a set of linguistic terms and (s_i,α) a 2-tuple. Then there is always a Δ^{-1} function such that for each 2-tuple it returns with its equivalent numerical value $\beta \in [0,g]$, and

$$\Delta^{-1}: S\times[-0.5,0.5) \longrightarrow [0,g]$$
$$\Delta^{-1}(S_i,\alpha) = i+\alpha = \beta \tag{5.2}$$

From Definitions 5.1, 5.2, and 5.3, it follows that the conversion of a linguistic term into a linguistic 2-tuple consists of adding a value 0 as symbolic translation as follows:

$$s_i \in S \Rightarrow (s_i,0) \tag{5.3}$$

The 2-tuple linguistic computational model operates on the previously defined 2-tuples without losing any available information and is defined on the bases on the following details:

(1) The negation operator of a 2-tuple: $Neg((s_i,\alpha)) = \Delta(g-(\Delta^{-1}(s_i,\alpha)))$;

(2) Comparison of 2-tuples: The linguistic information represented by 2-tuples is compared according to the lexicographic order as follows. Let (s_k,α_1) and (s_l,α_2) be two linguistic 2-tuples:

- If $k<l$, then (s_k,α_1) is smaller than (s_l,α_2).
- If $k=l$, then
  * If $\alpha_1=\alpha_2$, then (s_k,α_1) and (s_l,α_2) represent the same information.
  * If $\alpha_1<\alpha_2$, then (s_k,α_1) is smaller than (s_l,α_2).
  * If $\alpha_1>\alpha_2$, then (s_k,α_1) is greater than (s_l,α_2).

(3) Aggregation of 2-tuples.

Definition 5.4 (Herrera and Martinez, 2001a; Herrera and Martinez, 2000). Let $x=\{(r_1,\alpha),\ldots,(r_n,\alpha_n)\}$ be a set of 2-tuples. Then the 2-tuple arithmetic mean $\bar{x}^e$ is computed as follows:

$$\bar{x}^e = \Delta\left(\sum_{i=1}^{n}\frac{1}{n}\Delta^{-1}(r_i,\alpha_i)\right) = \Delta\left(\frac{1}{n}\sum_{i=1}^{n}\beta_i\right) \tag{5.4}$$

Definition 5.5 (weighted arithmetic average) (Herrera and Martinez, 2000). Let $x=\{(r_1,\alpha),\ldots,(r_n,\alpha_n)\}$ be a set of 2-tuples, and $\Omega=\{\omega_1,\ldots,\omega_m\}$, satisfying $\sum_{i=1}^{m}\omega_i=1$, the associated weights. The 2-tuple weighted arithmetic average $\overline{x}^{\omega}$ is defined as follows:

$$\overline{x}^{\omega}=\Delta(\sum_{i=1}^{n}\omega_i\Delta^{-1}(r_i,\alpha_i))=\Delta(\sum_{i=1}^{n}\omega_i\beta_i) \tag{5.5}$$

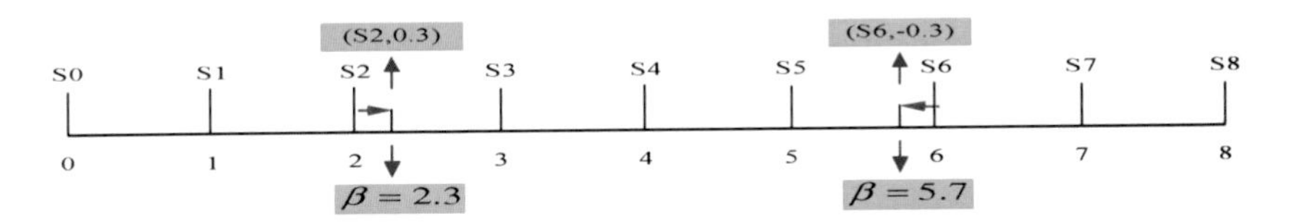

Fig. 5.3 A β value produced by aggregating linguistic information and equivalent 2-tuple (S_i,α_i)

Example 5.1. In Figure 5.3, let $S=\{S_0,S_1,\cdots,S_8\}$ be a set of linguistic terms, and $\beta=2.3$ produced by aggregating the linguistic information by using a chosen symbolic method. Then, according to Eqs. (5.2) and (5.3), $\Delta(2.3)=(S_2,0.3)$, the equivalent numerical value of $(S_2,0.3)$ is 2.3; and $(S_6,-0.3)>(S_2,0.3)$; $\text{Neg}(S_2,0.3)=\Delta(8-\Delta^{-1}(S_2,0.3))=\Delta(8-2.3)=(S_6,-0.3)$; $\text{Max}\{(S_6,-0.3),(S_2,0.3)\}=(S_6,-0.3)$; $\text{Min}\{(S_6,-0.3),(S_2,0.3)\}=(S_2,0.3)$.

5.2 Two-Tuple Linguistic Preference Relations

This section is based on (Gong and Liu, 2007c). For more details, please consult with this reference.

For a set $X=\{x_1,x_2,\ldots,x_n\}$ of decision alternatives, the preference information of pairwise comparisons with respect to a single criterion is represented numerically using a linguistic matrix $T=(s_{ij})_{n\times n}$ on the set $S=\{s_0,s_1,\ldots,s_g\}$ of linguistic terms. A 2-tuple $s_{ij}=(p_{ij},\alpha_{ij})$, where $p_{ij}\in S,\alpha_{ij}\in[-0.5,0.5)$, estimates the degree of linguistic preference of the decision alternative x_i over another alternative x_j. Particularly, $s_{ij} = (s_{\frac{g}{2}}, 0)$ indicates indifference between the alternatives x_i and x_j, $s_{ij} > (s_{\frac{g}{2}}, 0)$ that x_i is preferred to x_j, and $s_{ij} < (s_{\frac{g}{2}}, 0)$ that x_j is preferred to x_i.

Definition 5.6 (Herrera, Martinez and Sanchez, 2005). A linguistic matrix $T = (s_{ij})_{n \times n}$ is referred to as a 2-tuple linguistic preference relation (LPR), if the following hold true:

$$s_{ii} = (p_{ii}, \alpha_{ii}) = (S_{\frac{g}{2}}, 0)$$

$$\Delta(\Delta^{-1}(s_{ij}) + \Delta^{-1}(s_{ji})) = (s_g, 0) \text{ (complementary)}$$

where $s_{ij} = (p_{ij}, \alpha_{ij}), p_{ij} \in S, \alpha_{ij} \in [-0.5, 0.5), i, j \in N$.

Definition 5.7. A 2-tuple LPR $T = (s_{ij})_{n \times n}$ is said to be additively consistent, if the following conditions hold true:

1) $(p_{ii}, \alpha_{ii}) = (S_{\frac{g}{2}}, 0)$;

2) $\Delta(\Delta^{-1}(p_{ij}, \alpha_{ij}) + \Delta^{-1}(p_{ji}, \alpha_{ji})) = (S_g, 0)$; and

3) When $\Delta^{-1}(p_{ik}, \alpha_{ik}) \geq \frac{g}{2}, \Delta(\Delta^{-1}(p_{ij}, \alpha_{ij}) + \Delta^{-1}(p_{jk}, \alpha_{jk}) - g) = \Delta(\Delta^{-1}(p_{ik}, \alpha_{ik}) - \frac{g}{2})$; and when $\Delta^{-1}(p_{ik}, \alpha_{ik}) \leq \frac{g}{2}, \Delta(g - \Delta^{-1}(p_{ij}, \alpha_{ij}) - \Delta^{-1}(p_{jk}, \alpha_{jk})) = \Delta(\frac{g}{2} - \Delta^{-1}(p_{ik}, \alpha_{ik}))$.

Definition 5.8. A 2-tuple LPR $T = (s_{ij})_{n \times n}$ is said to have the property of restricted min-min transitivity, if $(p_{ij}, \alpha_{ij}) \geq (S_{\frac{g}{2}}, 0), (p_{jk}, \alpha_{jk}) \geq (S_{\frac{g}{2}}, 0) \Rightarrow (p_{ik}, \alpha_{ik}) \geq min\{(p_{ij}, \alpha_{ij}), (p_{jk}, \alpha_{jk})\}$, for all $i, j, k \in N, i \neq j \neq k$.

Definition 5.9. A 2-tuple LPR $T = (s_{ij})_{n \times n}$ is said to have the property of restricted max-max transitivity, if $(p_{ij}, \alpha_{ij}) \geq (S_{\frac{g}{2}}, 0), (p_{jk}, \alpha_{jk}) \geq (S_{\frac{g}{2}}, 0) \Rightarrow (p_{ik}, \alpha_{ik}) \geq max\{(p_{ij}, \alpha_{ij}), (p_{jk}, \alpha_{jk})\}$, for all $i, j, k \in N, i \neq j \neq k$.

Theorem 5.1. An additively consistent 2-tuple LPR also satisfies the property of restricted max-max transitivity.

Proof. Let $T=(s_{ij})_{n\times n}$ be an additively consistent 2-tuple LPR. When $\Delta^{-1}(p_{ik},\alpha_{ik})\geq\frac{g}{2}$, we have $\Delta(\Delta^{-1}(p_{ij},\alpha_{ij})+\Delta^{-1}(p_{jk},\alpha_{jk})-g)=\Delta(\Delta^{-1}(p_{ik},\alpha_{ik})-\frac{g}{2})$. And when $(p_{ij},\alpha_{ij})\geq(S_{\frac{g}{2}},0)$, we have $\Delta^{-1}(p_{ij},\alpha_{ij})\geq\frac{g}{2}$. Thus we obtain that

$$\Delta(\Delta^{-1}(p_{ik},\alpha_{ik})-\tfrac{g}{2})\geq\Delta(\Delta^{-1}(p_{jk},\alpha_{jk})-\tfrac{g}{2})$$

which in turn implies

$$\Delta^{-1}(p_{ik},\alpha_{ik})\geq\Delta^{-1}(p_{jk},\alpha_{jk}).$$

So, we have

$$(p_{ik},\alpha_{ik})\geq(p_{jk},\alpha_{jk}).$$

Similarly, when $(p_{jk},\alpha_{jk})\geq(S_{\frac{g}{2}},0)$, we can obtain $(p_{ik},\alpha_{ik})\geq(p_{ij},\alpha_{ij})$. QED.

According to the proof of Theorem 5.1, we can readily produce the following corollary.

Corollary 5.1. An additively consistent 2-tuple LPR also has the property of restricted min-min transitivity.

Definition 5.10. A 2-tuple LPR $T=(s_{ij})_{n\times n}$ is said to have the property of satisfactory consistency, if for all $i,j,k\in N, i\neq j\neq k$, the following hold true:

$$(p_{ij},\alpha_{ij})\geq(S_{\frac{g}{2}},0),(p_{jk},\alpha_{jk})\geq(S_{\frac{g}{2}},0)\Rightarrow(p_{ik},\alpha_{ik})\geq(S_{\frac{g}{2}},0);$$
$$(p_{ij},\alpha_{ij})\leq(S_{\frac{g}{2}},0),(p_{jk},\alpha_{jk})\leq(S_{\frac{g}{2}},0)\Rightarrow(p_{ik},\alpha_{ik})\leq(S_{\frac{g}{2}},0).$$

The equivalent to Definition 5.10 is the following Definition 5.11.

Definition 5.11. If a 2-tuple LPR $T=(s_{ij})_{n\times n}$ has the property of satisfactory consistency, then the corresponding preference of the decision alternatives $X=\{x_1,x_2,\cdots,x_n\}$ has the property of transitivity property. That is, there exists a priority chain $x_{u_1}\geq x_{u_2}\geq\cdots\geq x_{u_n}$ in the set $X=\{x_1,x_2,\cdots,x_n\}$, where x_{u_i} denotes the ith decision alternative in the priority chain, and $x_{u_i}\geq x_{u_j}$ represents that the decision alternative x_{u_i} is preferred (superior) to x_{u_j}. If there exists a circulation $x_{u_{i1}} \geq x_{u_{i2}} \geq \cdots \geq x_{u_{ik}} \geq x_{u_{i1}}$ among the decision alternatives, then the corresponding preference relation on the set $X = \{x_1, x_2, \cdots, x_n\}$ does not have any of the properties of transitivity; and the 2-tuple LPR $T = (s_{ij})_{n\times n}$ is said to be inconsistent.

Theorem 5.2. An additively consistent 2-tuple LPR has also the property of satisfactory consistency.

The proof is straightforward and is omitted. QED.

According to Definition 5.11 and Theorem 5.2, it can be seen that the property of satisfactory consistency is the minimum logical requirement and a fundamental principle of 2-tuple LPRs. It reflects the thinking characteristics of man. Therefore, it is very important to set up an easy-to-follow approach that can be used to judge whether a given 2-tuple LPR satisfies the property of satisfactory consistency. In the following, we introduce the definition of preference matrices.

Definition 5.12. Let **T**=$(p_{ij}, \alpha_{ij})_{n\times n}$ be a 2-tuple LPR. $Q = (q_{ij})_{n\times n}$ is referred to as a preference matrix of **T**, if

$$q_{ij} = \begin{cases} 1 & (p_{ij}, \alpha_{ij}) > (S_{\frac{g}{2}}, 0) \\ 0 & otherwise \end{cases}.$$

Theorem 5.3. Let $Q = (q_{ij})_{n\times n}$ be the preference matrix of **T**=$(p_{ij}, \alpha_{ij})_{n\times n}$, $\mathbf{Q}^i$ the ith sub-matrix of $Q = (q_{ij})_{n\times n}$, $i(i = 0, 1, \cdots, n-1, n)$. (That is, by deleting one 0 row vector and the corresponding same-numbered column vector from Q, a sub-matrix $\mathbf{Q}^1$ of $\mathbf{Q}^0$=**Q** is obtained; · · ·; by deleting one 0 row vector and the corresponding same-numbered column vector of $\mathbf{Q}^i$, a sub-matrix $\mathbf{Q}^{i+1}$ of $\mathbf{Q}^i$ is obtained; · · ·; by deleting one 0 row vector and the corresponding same-numbered column vector of $\mathbf{Q}^{n-1}$, a sub-matrix $\mathbf{Q}^{n-2}$ of $\mathbf{Q}^{n-1}$ is obtained, and $\mathbf{Q}^n = (0)$.) For all $i(i = 0, 1, \cdots, n-1, n)$, **T**= $(p_{ij}, \alpha_{ij})_{n\times n}$ is said to be satisfactorily consistent if and only if there is a 0 row vector in $\mathbf{Q}^i$.

Proof. Necessity. If **T**=$(p_{ij}, \alpha_{ij})_{n\times n}$ has the property of satisfactory consistency, suppose that we have a set $X = \{x_1, x_2, \cdots, x_n\}$ of decision alternatives such that the alternatives can be ranked in as priority chain $x_{u1} \geq x_{u2} \geq \cdots \geq x_{un}$, where

x_{ui} denotes the ith alternative in the priority chain. Because x_{un} is the most inferior decision alternative, it follows that $(p_{u_n j}, \alpha_{u_n j}) \le (S_{\frac{g}{2}}, 0)$, so that $q_{u_n j} = 0, j = 1, \cdots, n$. That is, the entries in the u_n-th row are all 0. Deleting the u_n-th row and the u_n-th column from Q, we obtain a sub-matrix $\mathbf{Q}^1$. At this time, the pairwise priority relations of the remaining decision alternatives are not changed. Thus $x_{u_{n-1}}$ becomes the most inferior among the remaining decision alternatives. Obviously, in $\mathbf{Q}^1$, the entries of the row represented by $x_{u_{n-1}}$ are all 0. By deleting the u_nth row and the u_nth column, the u_{n-1}th row and the u_{n-1}th column from **Q**, we obtain a submatrix $\mathbf{Q}^2$. Continuing this procedure, at the end we produce an $(n-1)$th sub-matrix as follows:

$$\mathbf{Q}^{n-1} = \begin{pmatrix} 0 & 1 \\ 0 & 0 \end{pmatrix} \text{or} \begin{pmatrix} 0 & 0 \\ 1 & 0 \end{pmatrix} \text{or} \begin{pmatrix} 0 & 0 \\ 0 & 0 \end{pmatrix}$$

In $\mathbf{Q}^{n-1}$, the 0 row vector is represented by x_{u_2}. Deleting the 0 row vector and the corresponding same-numbered column leads to a sub-matrix $Q^n = (0)$ so that the most superior alternative x_{u_1} is consequently produced.

Sufficiency. Let the entries of the u_nth row vector in **Q** be all 0. Then it can be readily seen that x_{u_n} is the most inferior decision alternative. Now by deleting the u_nth row and the u_nth column in **Q**, we obtain a sub-matrix $\mathbf{Q}^1$. Let the 0 row vector in $\mathbf{Q}^1$ be represented by $x_{u_{n-1}}$, then $x_{u_{n-1}}$ is superior to x_{u_n}. Continuing this procedure, at the very end we obtain the most superior alternative x_{u_1}. Thus we produce a priority chain $x_{u_1} \ge x_{u_2} \ge \cdots \ge x_{u_n}$, where x_{n_i} denotes the ith superior alternative in the set $X = \{x_1, x_2, \cdots, x_n\}$ of all decision alternatives. Therefore, **T** has the property of satisfactory consistency. QED.

As a matter of fact, based on Theorem 5.3, we can establish a priority algorithm for each satisfactorily consistent 2-tuple LPR as follows:

Step 1. Construct the preference matrix.
Step 2. Let $i = 0$.

Step 3. Search for the 0 row vector in the sub-matrix $\mathbf{Q}^i$. If such a 0 row exists, then the decision alternative represented this row is denoted $x_{u_{n-i}}$, and go to step 4. Otherwise go to step 5.

Step 4. Delete the 0 row in $\mathbf{Q}^i$ (if there are more than 1 such rows, then randomly select one such a 0 row) and the corresponding same-numbered column from $\mathbf{Q}^i$, and set $i = i + 1$. If $i = n$, then the decision alternative represented by this row is denoted x_{u_1} (That is, in this case, T has the property of satisfactory consistency). End. Otherwise, go to step 3.

Step 5. T is inconsistent. End

Theorem 5.4. The arithmetic average combination of additively consistent 2-tuple LPRs still has additive consistency.

Proof. Let $\mathbf{T}_k = \left(\left(p_{ij}^{(k)}, \alpha_{ij}^{(k)}\right)\right)_{n\times n}$ be the kth additively consistent 2-tuple LPRs, and λ_k the corresponding weight, satisfying $\sum_{k=1}^{l} \lambda_k = 1$, where $\lambda_k \geq 0, k = 1, 2, \cdots, l$. Then, we have

$$\sum_{k=1}^{l} \lambda_k (p_{ij}^{(k)}, \alpha_{ij}^{(k)}) + \sum_{k=1}^{l} \lambda_k (p_{jm}^{(k)}, \alpha_{jm}^{(k)}) + \sum_{k=1}^{l} \lambda_k (p_{mi}^{(k)}, \alpha_{mi}^{(k)})$$

$$= \sum_{k=1}^{l} (round(\lambda_k (p_{ij}^{(k)} + p_{jm}^{(k)} + p_{mi}^{(k)} + \alpha_{ij}^{(k)} + \alpha_{jm}^{(k)} + \alpha_{mi}^{(k)})), \lambda_k (p_{ij}^{(k)} + p_{jm}^{(k)} + p_{mi}^{(k)} + \alpha_{ij}^{(k)}$$

$$+ \alpha_{jm}^{(k)} + \alpha_{mi}^{(k)}) - round(\lambda_k (p_{ij}^{(k)} + p_{jm}^{(k)} + p_{mi}^{(k)} + \alpha_{ij}^{(k)} + \alpha_{jm}^{(k)} + \alpha_{mi}^{(k)}))) = \sum_{k=1}^{l} (\lambda_k S_{\frac{3g}{2}}, 0) = (S_{\frac{3g}{2}}, 0)$$

Similarly, we can prove $\sum_{k=1}^{l} \lambda_k (p_{ii}^{(k)}) = (S_{\frac{g}{2}}, 0)$, $\sum_{k=1}^{l} \lambda_k (p_{ij}^{(k)}) + \sum_{k=1}^{l} \lambda_k (p_{ji}^{(k)}) = (S_g, 0)$.

QED.

At this junction, we like to mention that the arithmetic average combination of satisfactorily consistent 2-tuple LPRs may not have the property of satisfactory consistency.

Example 5.2. Suppose that there are two decision makers who respectively provide their 2-tuple LPRs T_1 and T_2 on a set $X = \{x_1, x_2, x_3\}$ of decision alternatives as follows:

$$T_1 = \begin{pmatrix} (S_6,0) & (S_5,0) & (S_3,0) \\ (S_7,0) & (S_6,0) & (S_4,0) \\ (S_9,0) & (S_8,0) & (S_6,0) \end{pmatrix}; \; T_2 = \begin{pmatrix} (S_6,0) & (S_9,0) & (S_8,0) \\ (S_3,0) & (S_6,0) & (S_{10},0) \\ (S_4,0) & (S_2,0) & (S_6,0) \end{pmatrix}.$$

We also suppose that T_1 and T_2 are satisfactorily consistent 2-tuple LPRs, and that the two decision makers have the same weight. Then, the arithmetic average combination of T_1 and T_2 is

$$T = 0.5\begin{pmatrix} (S_6,0) & (S_5,0) & (S_3,0) \\ (S_7,0) & (S_6,0) & (S_4,0) \\ (S_9,0) & (S_8,0) & (S_6,0) \end{pmatrix} + 0.5\begin{pmatrix} (S_6,0) & (S_9,0) & (S_8,0) \\ (S_3,0) & (S_6,0) & (S_{10},0) \\ (S_4,0) & (S_2,0) & (S_6,0) \end{pmatrix}$$
$$= \begin{pmatrix} (S_6,0) & (S_7,0) & (S_6,-0.5) \\ (S_5,0) & (S_6,0) & (S_7,0) \\ (S_7,-0.5) & (S_5,0) & (S_6,0) \end{pmatrix}.$$

The preference matrix of T is

$$\begin{pmatrix} 0 & 1 & 0 \\ 0 & 0 & 1 \\ 1 & 0 & 0 \end{pmatrix}.$$

According to Theorem 5.3, T is not a satisfactorily consistent 2-tuple LPR.

5.3 Relationship between FPRs and 2-Tuple LPRs

The definition and properties of fuzzy preference relations (FPRs) have been given in Chapter 2. In this section, we discuss the relationship between FPRs and 2-tuple LPRs. For more details, please see (Gong and Liu, 2007c; Herrera, Martinez, and Sanchez, 2005).

Definition 5.13 (Herrera, Martinez, and Sanchez, 2005). Let $S = \{S_0, S_1, \cdots, S_g\}$ be a set of linguistic terms. The function τ, as defined in the following, transforms each numerical value $v \in [0,1]$ into a set of 2-tuple linguistic terms:

$$\tau : [0,1] \longrightarrow F(S)$$
$$\tau(v) = \{(S_0, \omega_0), (S_1, \omega_1), \cdots, (S_g, \omega_g)\} \tag{5.5}$$

$$\omega_i = \mu_{s_i}(v) = \begin{cases} 0 & \text{if } v \notin support(\mu_{s_i(x)}); \\ \dfrac{v - a_i}{b_i - a_i} & \text{if } a_i \le v \le b_i; \\ \dfrac{c_i - v}{c_i - b_i} & \text{if } b_i \le v \le c_i \end{cases} \tag{5.6}$$

where $S_i \in S, \omega_i \in [0,1], \mu_{S_i}(v)$ represents the membership degree of

$$Support(\mu_{s_i}(x)) = \{x \mid \mu_{s_i}(x) > 0, \quad x \in [0,1]\}.$$

Definition 5.14 (Herrera, Martinez and Sanchez, 2005). Let $\tau(v) = \{(S_0, \omega_0), \cdots, (S_g, \omega_g)\}$ be the transformed value of v by using Eq. (5.6). The function χ, as defined below, transforms the set $\tau(v)$ of 2-tuple linguistic terms into a representative β of 2-tuples.

$$\chi : F(S) \longrightarrow [0, g]$$
$$\chi(\tau(v)) = \chi\{(S_j, \omega_j), j = 0,1,\cdots, g\} = \sum_{j=0}^{g} j\omega_j \Big/ \sum_{j=0}^{g} \omega_j = \beta \tag{5.7}$$

From Definition 5.2, it follows that β can be transformed into its representative 2-tuple. For more details, see Figure 5.4.

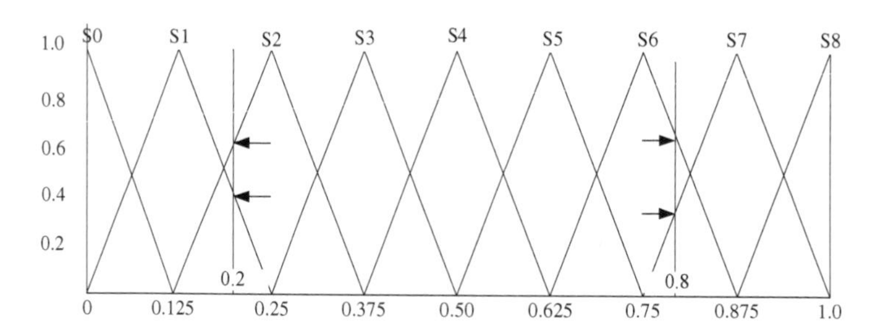

Fig. 5.4 The relationship between a numerical value and a set of 2-tuple linguistic terms

In Chapter 2, a necessary and sufficient condition is given for additively consistent FPRs, and it is also shown that each additively consistent FPR has the properties of satisfactory consistency, restricted max-max transitivity, and restricted max-min transitivity. In this section, we will show that the 2-tuple LPR of a given FPR as transformed by using Eqs. (5.2), (5.5)-(5.7) still has these properties.

Theorem 5.5. The transformed 2-tuple LPR of any chosen additively consistent FPR is still additively consistent.

Proof. Let $A = (a_{ij})_{n\times n}$ be an additively consistent FPR satisfying

$$a_{ii} = 0.5, a_{ij} + a_{ji} = 1, a_{ij} + a_{jk} + a_{ki} = 1.5,$$

and $S = \{S_0, \cdots, S_g\}$ a set of linguistic terms satisfying $S_i = (a_i, b_i, c_i)$, and

$$a_i = \tfrac{i-1}{g} (1 \le i \le g), b_i = \tfrac{i}{g}\ (0 \le i \le g), c_i = \tfrac{i+1}{g} (0 \le i \le g-1), a_0 = 0, c_g = 1$$

From Eq. (5.5), we have

$$\tau(a_{ii}) = \tau(0.5) = \{(S_0, 0), (S_1, 0), \cdots, (S_{\frac{g}{2}-1}, 0), (S_{\frac{g}{2}}, 1), (S_{\frac{g}{2}+1}, 0), \cdots, (S_g, 0)\}$$

Therefore, we have

$$\chi(\tau(0.5)) = \tfrac{g}{2}$$

That is, we have

$$\Delta^{-1}(\tfrac{g}{2}) = (S_{\frac{g}{2}}, 0)$$

Let $b_l \le a_{ij} \le c_l, a_{l+1} \le a_{ij} \le b_{l+1}$; $b_m \le a_{jk} \le c_m, a_{m+1} \le a_{jk} \le b_{m+1}$; $b_n \le a_{ki} \le c_n$, $a_{n+1} \le a_{ki} \le b_{n+1}$, $l, m, n \in \{0, 1, \cdots, g\}$. By applying Eq. (5.6), we conclude that

$$\omega_l = \frac{c_l - a_{ij}}{c_l - b_l}, \omega_{l+1} = \frac{a_{ij} - a_{l+1}}{b_{l+1} - a_{l+1}}, \omega_i = 0 \text{ hold for all } i \ne l, l+1;$$

$$\omega_m = \frac{c_m - a_{jk}}{c_m - b_m}, \omega_{m+1} = \frac{a_{jk} - a_{m+1}}{b_{m+1} - a_{m+1}}, \omega_i = 0 \text{ hold for all } i \ne m, m+1;$$

and

$$\omega_n = \frac{c_n - a_{ki}}{c_n - b_n}, \omega_{n+1} = \frac{a_{ki} - a_{n+1}}{b_{n+1} - a_{n+1}}, \omega_i = 0 \text{ hold for all } i \ne n, n+1.$$

Therefore, we have

$$\chi(\tau(a_{ij})) = lg(\tfrac{l+1}{g} - a_{ij}) + (l+1)g(a_{ij} - \tfrac{l}{g})$$

$$\chi(\tau(a_{jk})) = mg(\tfrac{m+1}{g} - a_{jk}) + (m+1)g(a_{jk} - \tfrac{l}{g}); \chi(\tau(a_{ki}))$$

$$= ng(\tfrac{n+1}{g} - a_{ki}) + (n+1)g(a_{ki} - \tfrac{l}{g})$$

So, the following holds true:

$$\chi(\tau(a_{ij})) + \chi(\tau(a_{jk})) + \chi(\tau(a_{ki})) = g(a_{ij} + a_{jk} + a_{ki}) = 1.5g$$

which in turn implies

$$\Delta^{-1}(\chi(\tau(a_{ij})) + \chi(\tau(b_{ij})) + \chi(\tau(c_{ij}))) = (S_{1.5g}, 0).$$

Similarly, we can prove that $\Delta^{-1}(\chi(\tau(a_{ij})) + \chi(\tau(a_{ji})) = (S_g, 0)$. QED.

Theorem 5.6. The transformed 2-tuple LPR of a satisfactorily consistent (restricted max-max transitive, restricted max-min transitive) FPR is also additively consistent.

The proof is very similar to that of Theorem 5.5, so it is omitted.

According to Theorem 5.5 and 5.6, it can be seen that the transformed 2-tuple LPR of a given FPR maintains the integrity of original information.

5.4 A Numerical Example

Routine planning is a vitally important matter of airlines. For routing choices, consideration needs to be given to the following five important factors: short-term average load factor, competitiveness, influence of flight route network, the mid- and long-term market demand, and ticket cost. An air transportation company invites three experts (e_1, e_2, e_3) to evaluate four airlines named $x_i, i = 1, 2, 3, 4$, by comprehensively evaluating these five factors.

Let us suppose that the corresponding weights of these experts are respectively 0.4, 0.3, and 0.3, and the set of linguistic terms is $S = \{N, VL, L, M, H, VH, P\}$. Additionally, we suppose that the first expert presents his fuzzy preference relation P^1, the second and the third experts present their 2-tuple linguistic preference relations P^2 and P^3, where

$$\mathbf{P}^1 = \begin{pmatrix} 0.5 & 0.7 & 0.6 & 0.8 \\ 0.3 & 0.5 & 0.4 & 0.6 \\ 0.4 & 0.6 & 0.5 & 0.7 \\ 0.2 & 0.4 & 0.3 & 0.5 \end{pmatrix}; \mathbf{P}^2 = \begin{pmatrix} M & H & M & VH \\ L & M & L & H \\ M & H & M & VH \\ VL & L & VL & M \end{pmatrix};$$

$$\mathbf{P}^3 = \begin{pmatrix} M & VH & VH & H \\ VL & M & M & L \\ VL & M & M & L \\ L & H & H & M \end{pmatrix}$$

Step 1: According to Eqs. (5.1), (5.2), (5.5), and (5.7), we transform these matrices into 2-tuple LPRs as follows:

$$\mathbf{P}^1 = \begin{pmatrix} (S_3,0) & (S_4,0.2) & (S_4,-0.4) & (S_5,-0.2) \\ (S_2,-0.2) & (S_3,0) & (S_2,0.4) & (S_4,-0.4) \\ (S_2,0.4) & (S_4,-0.4) & (S_3,0) & (S_4,0.2) \\ (S_1,0.2) & (S_2,0.4) & (S_2,-0.2) & (S_3,0) \end{pmatrix}$$

$$\mathbf{P}^2 = \begin{pmatrix} (S_3,0) & (S_4,0) & (S_3,0) & (S_5,0) \\ (S_2,0) & (S_3,0) & (S_2,0) & (S_4,0) \\ (S_3,0) & (S_4,0) & (S_3,0) & (S_5,0) \\ (S_1,0) & (S_2,0) & (S_1,0) & (S_3,0) \end{pmatrix}$$

$$\mathbf{P}^3 = \begin{pmatrix} (S_3,0) & (S_5,0) & (S_5,0) & (S_4,0) \\ (S_1,0) & (S_3,0) & (S_3,0) & (S_2,0) \\ (S_1,0) & (S_3,0) & (S_3,0) & (S_2,0) \\ (S_2,0) & (S_4,0) & (S_4,0) & (S_3,0) \end{pmatrix}$$

Step 2: By using Eq.(5.7), we aggregate $\mathbf{P}^1$, $\mathbf{P}^2$, and $\mathbf{P}^3$ into the following collective LPR $\mathbf{P}$ of 2-tuples.

$$\mathbf{P} = \begin{pmatrix} (S_3,0) & (S_4,0.38) & (S_4,-0.16) & (S_5,-0.38) \\ (S_2,-0.38) & (S_3,0) & (S_2,0.46) & (S_3,0.24) \\ (S_2,0.16) & (S_4,-0.46) & (S_3,0) & (S_4,-0.22) \\ (S_1,0.38) & (S_3,-0.24) & (S_2,0.22) & (S_3,0) \end{pmatrix}$$

Step 3.1: By applying Theorem 5.3, we construct the preference matrix of **P** as follows:

$$\mathbf{R} = \begin{pmatrix} 0 & 1 & 1 & 1 \\ 0 & 0 & 0 & 1 \\ 0 & 1 & 0 & 1 \\ 0 & 0 & 0 & 0 \end{pmatrix}.$$

Step 3.2. Let $\mathbf{R}^0 = \mathbf{R}$.

Step 3.3. Search for a 0 row. Obviously, the fourth row in **R** is 0. This means that x_4 is the most inferior alternative.

Step 3.4. Delete the fourth row and the fourth column from *R* so that we obtain the following sub-matrix

$$\mathrm{R}^1 = \begin{pmatrix} 0 & 1 & 1 \\ 0 & 0 & 0 \\ 0 & 1 & 0 \end{pmatrix}.$$

Step 3.5. Search for a 0 row in R^1. Obviously, the second row in R^1 is 0. And this row is also the second row of R. So, x_2 is the second most inferior decision alternative.

Similarly, we obtain the sub-matrices

$$\mathrm{R}^2 = \begin{pmatrix} 0 & 1 \\ 0 & 0 \end{pmatrix} \text{ and } \mathrm{R}^3 = (0),$$

which means that x_3 is the second most optimal decision alternative, while x_1 is the most optimal alternative.

In light of Theorem 5.3 and the corresponding algorithms, P is a satisfactorily consistent 2-tuple LPR and the ranking of four airlines is $x_1 \succ x_3 \succ x_2 \succ x_4$.

5.5 Conclusions

In this chapter, we studied the proprieties of two-tuple LPRs, and discussed the inherent relations among these proprieties. Meanwhile, we developed a priority algorithm based on satisfactorily consistent two-tuple LPRs. By using the transformation relations between two-tuple linguistic values and numerical numbers, we showed by establishing two theorems that the two-tuple LPR as obtained by transformation keeps the properties of consistency and other proprieties so that the available information is guaranteed to be intact and genuine.

The concept of two-tuple linguistic evaluations combines the merits of numerical information and linguistic qualitative information. It provides new ideas to the ways on how different preferences of decision makers can be aggregated. Techniques of decision making based on linguistic evaluations have been widely employed in such fields as investment decisions, personnel management, project evaluations, and social systems engineering.

Chapter 6
Preference Relations of Trapezoidal Fuzzy Numbers

As described in Chapter 5, qualitative information, such as that conveyed in outstanding, good and poor, is often used to assess qualitative variables, human moral, the performance of certain equipment, etc. In any fuzzy decision making, the available qualitative information is generally characterized by trapezoidal fuzzy numbers, triangular fuzzy numbers, and even natural linguistic terms (Fan, and Jiang, 2003, 2004; Xu, 2004e,2005b, 2005c;; Chen and Fan, 2004; Hou and Wu, 2005a, 2005b; Jiang and Fan, 2003; Chen and Fan, 2004; Herrera, 2001; Herrera, Herrera, and Martinez, 2000; Herrera and Martinez, 2000; 2001a; 2001b; Delgado, Herrera, and Herrera, 2002; Herrera, Martinez, and Sanchez, 2005; Wang and Fan, 2003; You, Fan, and Li, 2005;Gong, 2007; Gong and Liu, 2007c, Hou and Wu,2005b).

In fact, natural linguistics is generally transformed into triangular or trapezoidal fuzzy numbers, while the concept of trapezoidal fuzzy numbers is a generalization of that of triangular fuzzy numbers. In consequence, the research on preference relations of trapezoidal fuzzy numbers is of the practical theoretical and practical significance.

The development of the operational laws of trapezoidal fuzzy numbers still represents an unsettled problem. Additionally, when aggregating preferences of trapezoidal fuzzy numbers that appear in a group decision making situation, the range of the involved interval may be gradually exaggerated with the number of arithmetic computational steps. That may cause distortion of the available information in terms of decision making. Herrera et al. (Herrera, Herrera, and Martinez, 2000; Herrera, Martinez, and Sanchez, 2005) showed that crisp numbers, intervals, and triangular fuzzy numbers can all be transformed into the two-tuple linguistics by employing a transformation function. In this chapter, we will generalize this transformation function to the case of trapezoidal fuzzy numbers, and study the aggregation and the priority problem of the preference relations of trapezoidal fuzzy numbers.

6.1 Relationship between Preference Relations of Trapezoidal Fuzzy Numbers and 2-Tuple LPRs

A fuzzy number $\tilde{A}$ is referred to be as a trapezoidal fuzzy number, if its degree of membership $\mu_{\tilde{A}} : R \mapsto [0,1]$ satisfies the following:

Z. Gong et al.: Uncertain Fuzzy Preference Relations, STUDFUZZ 281, pp. 93–102.
springerlink.com © Springer-Verlag Berlin Heidelberg 2013

$$\mu_{\tilde{A}} = \begin{cases} (x-a)/(b-a), & a \le x \le b \\ 1, & b \le x \le c \\ (x-d)/(c-d), & c \le x \le d \\ 0, & \text{otherwise} \end{cases}$$

A trapezoidal fuzzy number can be denoted by using an ordered quadruple (a, b, c, d). In this book, we assume that all trapezoidal fuzzy numbers satisfy the condition that $0 < a \le b \le c \le d \le 1$.

Definition 6.1 (Hou and Wu. 2005b; Gong, 2007)**.** Let $\boldsymbol{R} = (r_{ij})_{n \times n}$ be a preference relation. Then $\boldsymbol{R}$ is referred to as preference relation of trapezoidal fuzzy numbers, if the following hold true:

$$r_{ij} = (a_{ij}, b_{ij}, c_{ij}, d_{ij}), 0 < a_{ij} \le b_{ij} \le c_{ij} \le d_{ij} \le 1, \ i, j = 1, \cdots, n,$$
$$a_{ii} = b_{ii} = c_{ii} = d_{ii} = 0.5 \text{ ; and}$$
$$a_{ij} + d_{ji} = b_{ij} + c_{ji} = c_{ij} + b_{ji} = d_{ij} + a_{ji} = 1 \text{ (complementary property).}$$

The following transformation functions establish the relationship between trapezoidal fuzzy numbers and 2-tuple linguistic terms.

Definition 6.2 (Herrera, Martinez, and Sanchez, 2005; Gong, 2007)**.** Let I be a trapezoidal fuzzy number, and $S_T = \{S_0, S_1, \cdots, S_g\}$ a set of linguistic terms. Then the function τ_{IS_T} transforms I into a fuzzy set in S_T :

$$\tau : [0,1] \longrightarrow F(S_T)$$
$$\tau(I) = \{(S_k, \omega_k)/k \in \{0, \cdots, g\}\} \tag{6.1}$$
$$\omega_k = \max_y min\{\mu_I(y), \mu_{S_k}(y)\} \tag{6.2}$$

where $F(S_T)$ is the set of all fuzzy sets defined on S_T, and $\mu_I(.)$ and $\mu_{S_k}(.)$ are the membership functions associated with the fuzzy number I and the terms in S_k, respectively.

Definition 6.3 (Herrera, Martinez, and Sanchez, 2005; Gong, 2007)**.** Let $\tau(I) = \{(S_0, \omega_0), \cdots, (S_g, \omega_g)\}$ be a fuzzy set that represents a numerical value I

over the set $S=\{S_0,S_1,\cdots,S_g\}$ of linguistic terms. Then a numerical value by means of the below function χ that represents the information of the fuzzy set can be obtained as follows:

$$\chi : F(S_T) \longrightarrow [0,g]$$
$$\chi\tau(I)=\chi(F(S_T))=\chi\{(S_j,\omega_j), j=0,1,\cdots,g\}=\sum_{j=0}^{g} j\omega_j(\sum_{j=0}^{g}\omega_j)^{-1}=\beta \quad (6.3)$$

Therefore, we can transform β of eq. (6.3) into it's representative 2-tuple linguistic value.

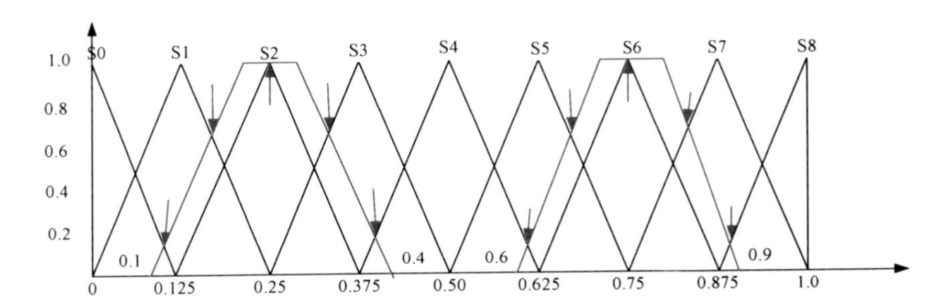

Fig. 6.1 The Relation between the Triangular Fuzzy Number and 2-tuple Linguistic Value

The following examples show that the transformed 2-tuple linguistic terms of complementary trapezoidal fuzzy numbers by obtained by employing eqs. (6.1) - (6.3) maintain the complementary property.

Example 6.1. In reference to Figure 6.1, let $S=\{S_0,S_1,\cdots,S_8\}$ be a set of linguistic terms. By applying eqs. (6.1) - (6.3), we can produce

$$\tau(0.1,0.2,0.3,0.4)=\{(S_0,0.11),(S_1,0.67),(S_2,1),(S_3,0.67),(S_4,0.11),(S_5,0)\cdots,(S_8,0)\},$$

and

$$\chi\tau(0.1,0.2,0.3,0.4)=(1\cdot 0.67+2\cdot 1+3\cdot 0.67+4\cdot 0.11)\ (0.11+0.67+1+0.67+0.11)^{-1}=2.$$

Because $\Delta(2)=(S_2,0)$, the transformation value of (0.1,0.2,0.3,0.4) is $(S_2,0)$. Similarly, the transformation value of (0.6,0.7,0.8,0.9) is $(S_6,0)$. So $\Delta(\Delta^{-1}(S_2,0)+\ \Delta^{-1}(S_6,0))=(S_8,0)$. The trapezoidal fuzzy numbers (0.1,0.2,0.3,0.4) and (0.6,0.7,0.8,0.9) are complementary; and the transformed 2-tuple linguistic terms also satisfy the complementary property.

With regard to the complementary property of preference relations of trapezoidal fuzzy numbers, the following result can be explained as follows: The transformation functions (Definitions 6.2 - 6.3) between trapezoidal fuzzy numbers and 2-tuples are effective.

Theorem 6.1. The 2-tuple LPR that is transformed from a preference relation of trapezoidal fuzzy numbers by using eqs. (6.1) - (6.3) still has the complementary property.

Proof. Let $S=\{S_0,\cdots,S_g\}$ be a set of linguistic terms, where

$$S_k=(a_k,b_k,c_k), a_k=\tfrac{k-1}{g}(1\le k\le g), b_k=\tfrac{k}{g}(0\le k\le g),\ c_k=\tfrac{k+1}{g}(0\le k\le g-1), a_0=0, c_g=1;$$

and $R=(a_{ij})_{n\times n}$ a preference relation of trapezoidal fuzzy numbers, where

$$a_{ii}=0.5, b_{ii}=0.5, c_{ii}=0.5, d_{ii}=0.5;\ a_{ij}+d_{ji}=1, b_{ij}+c_{ji}=1, c_{ij}+b_{ji}=1, d_{ij}+a_{ji}=1.$$

By employing eq. (6.2), we have

$$\tau(a_{ii})=\tau(0.5,0.5,0.5,0.5)=\{(S_0,0),(S_1,0),\cdots,\ (S_{\frac{g}{2}-1},0),(S_{\frac{g}{2}},1),(S_{\frac{g}{2}+1},0),\cdots,(S_g,0)\}.$$

so that the following holds true:

$$\chi(\tau(0.5,0.5,0.5,0.5))=\tfrac{g}{2}.$$

That is, we have $\Delta(\tfrac{g}{2})=(S_{g/2},0)$. Let $a_{ij}\in[b_l,c_l], b_{ij}\in[b_L,c_L], c_{ij}\in[a_m,b_m], d_{ij}\in[a_M,b_M]$. Then the following is true:

$$a_{ji}\in[b_{g-M},c_{g-M}], b_{ji}\in[b_{g-m},c_{g-m}], c_{ji}\in[a_{g-L},b_{g-L}], d_{ji}\in[a_{g-l},b_{g-l}],$$

where $j\ge i, l\le L\le m\le M$. Now, eq. (6.2) implies

$$\tau(a_{ij},b_{ij},c_{ij},d_{ij})=\{(S_0,\omega_0),\cdots,(S_{l-1},\omega_{l-1}),(S_l,\omega_l),\cdots,(S_L,\omega_L),$$
$$(S_{L+1},\omega_{L+1}),\cdots,(S_{m-1},\omega_{m-1}),(S_m,\omega_m),\cdots,(S_M,\omega_M),(S_{M+1},\omega_{M+1}),\cdots,(S_g,\omega_g)\}$$

where

$$\omega_0=\cdots=\omega_{l-1}=\omega_{M+1}=\cdots=\omega_g=0; \omega_{L+1}=\cdots=\omega_{m-1}=1;$$

$\omega_k = \frac{k+1-a_{ij}g}{1+g(b_{ij}-a_{ij})}$ is the ordinate of the intersection point of the line $y = -gx + k + 1$ and the line $y = \frac{x-a_{ij}}{b_{ij}-a_{ij}}$, for $k = l, \cdots, L$; and $\omega_K = \frac{K-1-d_{ij}g}{g(c_{ij}-d_{ij})-1}$ the ordinate of the intersection point of the line $y = gx - K + 1$ and the line $y = \frac{x-d_{ij}}{c_{ij}-d_{ij}}$, $K = m, \cdots, M$.

For

$$\tau(a_{ji}, b_{ji}, c_{ji}, d_{ji}) = \{(S_0, \omega_0'), \cdots, (S_{g-M}, \omega_{g-M}'), \cdots, (S_{g-m}, \omega_{g-m}'),$$
$$(S_{g-m+1}, \omega_{g-m+1}'), \cdots, (S_{g-L-1}, \omega_{g-L-1}'), (S_{g-L}, \omega_{g-L}'), \cdots, (S_{g-l}, \omega_{g-l}'), \cdots, (S_g, \omega_g')\}$$

where $\omega_0' = \cdots = \omega_{g-M-1}' = \omega_{g-l+1}' = \cdots = \omega_g' = 0;$ $\omega_{g-m+1}' = \cdots = \omega_{g-L-1}' = 1,$ $\omega_K' = \frac{K+1-a_{ji}g}{1+g(b_{ji}-a_{ji})}$, the ordinate of the intersection point of the line $y = -gx + K + 1$ and the line $y = \frac{x-a_{ji}}{b_{ji}-a_{ji}}$, for $K = g - M, \cdots, g - m$; $\omega_k' = \frac{k-1-d_{ji}g}{g(c_{ji}-d_{ji})-1}$ the ordinate of the intersection point of the line $y = gx - k + 1$ and the line $y = \frac{x-d_{ji}}{c_{ji}-d_{ji}}, k = g - L, \cdots, g - l$.

Thus we have

$$\begin{aligned}
\beta_1 &= \chi\tau(a_{ij}, b_{ij}, c_{ij}, d_{ij}) = \sum_{j=0}^{g} j\omega_j \left(\sum_{j=0}^{g} \omega_j\right)^{-1} \\
&= [i\omega_i + j\omega_j + (i+1)\omega_{i+1} + \cdots + (j-1)\omega_{j-1}](\omega_i + \cdots + \omega_j)^{-1} \\
&= [\sum_{k=l}^{L} k\frac{k+1-a_{ij}g}{1+g(b_{ij}-a_{ij})} + (L+1) + \cdots + (m-1) + \sum_{K=m}^{M} K\frac{K-1-d_{ij}g}{g(c_{ij}-d_{ij})-1}]x^{-1}
\end{aligned}$$

$$\begin{aligned}
\beta_2 &= \chi\tau(a_{ji}, b_{ji}, c_{ji}, d_{ji}) = \sum_{j=0}^{g} j\omega_j' \left(\sum_{j=0}^{g} \omega_j'\right)^{-1} \\
&= [(g-j)\omega_{g-j}' + (g-i)\omega_{g-i}' + (g-j+1) + \cdots + (g-i-1)][\omega_{g-j}' + \cdots + \omega_{g-i}']^{-1} \\
&= [(g-j)g(c_{g-j} - a_{ji}^{-}) + (g-j+1) + \cdots + (g-i-1) + (g-i)g(a_{ji}^{+} - a_{g-i})] \\
&\quad [g(c_{g-j} - a_{ji}^{-}) + j - i - 1 + g(a_{ji}^{+} - a_{g-i})]^{-1} \\
&= [\sum_{K=g-M}^{g-m} K\frac{K+1-a_{ji}g}{1+g(b_{ji}-a_{ji})} + (g-m+1) + \cdots + (g-L-1) + \sum_{k=g-L}^{g-l} k\frac{k-1-d_{ji}g}{g(c_{ji}-d_{ji})-1}]y^{-1}
\end{aligned}$$

Because

$$\begin{aligned}
&\sum_{k=l}^{L} \frac{k+1-a_{ij}g}{1+g(b_{ij}-a_{ij})} + m - L - 1 + \sum_{K=m}^{M} \frac{K-1-d_{ij}g}{g(c_{ij}-d_{ij})-1} \\
&= \sum_{K=g-M}^{g-m} \frac{K+1-a_{ji}g}{1+g(b_{ji}-a_{ji})} + m - L - 1 + \sum_{k=g-L}^{g-l} \frac{k-1-d_{ji}g}{g(c_{ji}-d_{ji})-1}
\end{aligned}$$

where

$$x=\sum_{k=l}^{L}\frac{k+1-a_{ij}g}{1+g(b_{ij}-a_{ij})}+m-L-1+\sum_{K=m}^{M}\frac{K-1-d_{ij}g}{g(c_{ij}-d_{ij})-1};$$

$$y=\sum_{K=g-M}^{g-m}\frac{K+1-a_{ji}g}{1+g(b_{ji}-a_{ji})}+m-L-1+\sum_{k=g-L}^{g-l}\frac{k-1-d_{ji}g}{g(c_{ji}-d_{ji})-1}.$$

we have readily that $x=y$.

It is readily to see that $\beta_1+\beta_2=g$. Thus $\Delta(\beta_1+\beta_2)=(S_g,0)$. That is, the transformed 2-tuple LPR has the complementary property.

6.2 Aggregation of Preference Relations of Trapezoidal Fuzzy Numbers

According to the conclusions of the previous analysis, the priority and aggregation of preference relations of trapezoidal fuzzy numbers can be transformed into those of the equivalent 2-tuple LPRs. So, the procedure of aggregating preference relations of trapezoidal fuzzy numbers is summarized as follows:

Step 1: Transform the given preference relation of trapezoidal fuzzy numbers into a 2-tuple LPR.

By using the transformation formulas, as given in eqs. (5.2), (6.1) - (6.3), we transform the given preference relation $(R_{ij}^k)_{n\times n}$ of trapezoidal fuzzy numbers into the 2-tuple LPR $(P_{ij}^k)_{n\times n}=(S_{ij}^k,\alpha_{ij}^k)_{n\times n}$; and we also transform the weights ω_k of the k -th decision maker into the 2-tuple linguistic terms $(f_k,\alpha_k), k=1,\cdots,m, f_k\in S$.

Step 2: Aggregate the transformed 2-tuple LPR. The collective preference relation $P=(p_{ij})_{n\times n}$ can be obtained by using the max-min operator in eq. (6.1):

$$P_{ij}=(S_{ij},\alpha_{ij})=\Delta(\frac{\sum_{k=1}^{m}\Delta^{-1}(S_{ij}^k,\alpha_{ij}^k).\Delta^{-1}(f_k,\alpha_k)}{\sum_{k=1}^{m}\Delta^{-1}(f_k,\alpha_k)}),i,j\in n \tag{6.4}$$

Step 3: Calculate the degree of how much a decision alternative is superior to the other decision alternatives.

Let the degree of how much a decision alternative x_i is superior to another decision alternative x_j be t_i, for $j=1,\cdots,n$. Then, we use the LOWA operator (linguistic order weighted operator) (Chiclana, Herrera, Herrera-Viedma, et al. 2003) to calculate the specific value of t_i, $j=1,\cdots,n$, where

$$t_i = \phi_Q(p_{i1},\cdots,p_{in}) = \Delta(\sum_{k=1}^{n} \omega_k^* \Delta^{-1}(p'_{ik})) \tag{6.5}$$

p'_{ik} is the kth largest label in the set $\{p_{i1},\cdots,p_{in}\}$.

The weight vector $\omega^* = (\omega_1^*,\cdots,\omega_k^*,\cdots,\omega_n^*)^T, k=1,\cdots,n$ can be given by (Filev and Yager, 1995):

$$Max \quad E(w) = -\sum_{i=1}^{n} w_i ln\omega_i \tag{6.6}$$

$$\text{s.t.} \begin{cases} orness(\omega) = \dfrac{1}{n-1}\sum_{i=1}^{n}(n-i)\omega_i \\ \sum_{i=1}^{n}\omega_i = 1,\ \omega_i \in [0,1],\ i=1,\cdots,n. \end{cases} \tag{6.7}$$

where the entropy $E(\omega)$ is being described as the degree of how well the decision making information is aggregated. The bigger the entropy $E(\omega)$ value is, the fuller the information is utilized. Let $orness(\omega) \in [0,1]$ be the degree of the decision maker's subjective attitude (either optimistic or pessimistic). The closer the distance between the $orness(\omega)$ value and 1 is, the more optimistic of the decision maker feels; the closer the distance between the $orness(\omega)$ value and 0 is, the more pessimistic of the decision maker is. The value of $orness(\omega)$ is determined by

$$\omega = (\omega_1,\cdots,\omega_k,\cdots,\omega_n)^T, \quad \omega_k = Q(k/n) - Q(k-1/n), k=1,\cdots,n$$

where $Q(r)$ stands for a fuzzy linguistic quantifier used to represent the fuzzy majority over dimension such that

$$Q(r) = \begin{cases} 0, & r < a, \\ (r-a)/(b-a), & a \le r \le b, \\ 1, & r > b. \end{cases}$$

Some linguistic fuzzy quantifiers are typified by such terms as "most ", "at least half", and "as many as possible ". The parameters (a,b) are $(0.3,0.8)$, $(0,0.5)$, and $(0.5,1)$, respectively.

For example, let $n=4$. By using the linguistic fuzzy quantifier "most" with the pair $(0.3,0.8)$, we obtain $\omega=(0,0.4,0.5,0.1)^T$, and $orness(\omega)=\frac{1}{3}\sum_{i=1}^{4}(4-i)\omega_i=0.4333$. That means that the decision maker has a neutral stand. Thus, we have $\omega^*=(0.1932,0.2269,0.2667,0.3133)^T$. By using the linguistic fuzzy quantifier "most" with the pair $(0,0.5)$, we obtain $\omega=(0.5,0.5,0,0)^T$ and orness(ω)=0.8333. That means that the decision maker has an optimistic stand. Thus, we have $\omega^*=(0.6478,0.2355,0.0856,0.0311)^T$. By using the linguistic fuzzy quantifier "most" with the pair $(0.5,1)$, $(a,b)=(0.5,1)$, we obtain $\omega=(0,0,0.5,0.5)^T$ and orness(ω)=0.1667. That means that the decision maker has a pessimistic stand. Thus, we have $\omega^*=(0.0311,0.0856,0.2355,0.6478)^T$.

Step 4: Rank the decision alternatives.

6.3 A Numerical Example

Suppose that there are four site selections $X=\{x_1,x_2,x_3,x_4\}$ in an investment project, and that the set of linguistic terms is $S=\{S_0,S_1,\cdots,S_8\}$. Let the set of decision makers be $E=\{e_1,e_2,e_3,e_4\}$, and the weights of all these decision makers be VL, VH, L and VL, respectively. Then the preference relations of trapezoidal fuzzy numbers as provided by the decision makers are listed as follows:

$$R_1=\begin{pmatrix}(0.5,0.5,0.5,0.5) & (0.5,0.6,0.7,0.8) & (0.6,0.7,0.8,0.9) & (0.5,0.6,0.6,0.7)\\ (0.2,0.3,0.4,0.5) & (0.5,0.5,0.5,0.5) & (0.1,0.2,0.2,0.4) & (0.2,0.3,0.4,0.5)\\ (0.1,0.2,0.3,0.4) & (0.6,0.8,0.8,0.9) & (0.5,0.5,0.5,0.5) & (0.7,0.8,0.8,0.9)\\ (0.3,0.4,0.4,0.5) & (0.5,0.6,0.7,0.8) & (0.1,0.2,0.2,0.3) & (0.5,0.5,0.5,0.5)\end{pmatrix}$$

$$R_2=\begin{pmatrix}(0.5,0.5,0.5,0.5) & (0.1,0.2,0.2,0.4) & (0.6,0.7,0.8,0.9) & (0.1,0.3,0.3,0.4)\\ (0.6,0.8,0.8,0.9) & (0.5,0.5,0.5,0.5) & (0.5,0.6,0.7,0.8) & (0.2,0.3,0.4,0.5)\\ (0.1,0.2,0.3,0.4) & (0.2,0.3,0.4,0.5) & (0.5,0.5,0.5,0.5) & (0.1,0.2,0.2,0.3)\\ (0.6,0.7,0.7,0.9) & (0.5,0.6,0.7,0.8) & (0.7,0.8,0.8,0.9) & (0.5,0.5,0.5,0.5)\end{pmatrix}$$

$$R_3=\begin{pmatrix}(0.5,0.5,0.5,0.5) & (0.2,0.3,0.4,0.5) & (0.6,0.7,0.8,0.9) & (0.3,0.4,0.4,0.5)\\ (0.5,0.6,0.7,0.8) & (0.5,0.5,0.5,0.5) & (0.5,0.6,0.7,0.8) & (0.5,0.5,0.5,0.5)\\ (0.1,0.2,0.3,0.4) & (0.2,0.3,0.4,0.5) & (0.5,0.5,0.5,0.5) & (0.1,0.2,0.2,0.3)\\ (0.5,0.6,0.6,0.7) & (0.5,0.5,0.5,0.5) & (0.7,0.8,0.8,0.9) & (0.5,0.5,0.5,0.5)\end{pmatrix}$$

Step 1: The transformed 2-tuple linguistic terms of the decision makers' weights are $(S_6,0),(S_3,0),(S_2,0)$. By utilizing the transformation formulas in eqs. (5.2), (6.1) - (6.3), we obtain the transformation matrices R_1, R_2 and R_3 to the 2-tuple linguistic preference relations as follows,

$$P_1=\begin{pmatrix} (S_4,0) & (S_5,0.26) & (S_6,0) & (S_5,-0.125) \\ (S_3,-0.26) & (S_4,0) & (S_2,-0.13) & (S_3,-0.26) \\ (S_2,0) & (S_6,0.13) & (S_4,0) & (S_6,0.375) \\ (S_3,0.125) & (S_5,0.26) & (S_2,-0.375) & (S_4,0) \end{pmatrix}$$

$$P_2=\begin{pmatrix} (S_4,0) & (S_2,-0.13) & (S_6,0) & (S_2,0.13) \\ (S_6,0.13) & (S_4,0) & (S_5,0.26) & (S_3,-0.26) \\ (S_2,0) & (S_3,-0.26) & (S_4,0) & (S_2,-0.375) \\ (S_6,-0.13) & (S_5,0.26) & (S_6,0.375) & (S_4,0) \end{pmatrix}$$

$$P_3=\begin{pmatrix} (S_4,0) & (S_3,-0.26) & (S_6,0) & (S_3,0.125) \\ (S_5,0.26) & (S_4,0) & (S_5,0.26) & (S_4,0) \\ (S_2,0) & (S_3,-0.26) & (S_4,0) & (S_2,-0.375) \\ (S_5,-0.125) & (S_4,0) & (S_6,0.375) & (S_4,0) \end{pmatrix}$$

Step 2: By applying eq.(6.4), the collective LPR is obtained as follows:

$$P=\begin{pmatrix} (S_4,0) & (S_4,-0.13) & (S_6,0) & (S_4,-0.19) \\ (S_4,0.13) & (S_4,0) & (S_3,0.41) & (S_3,-0.03) \\ (S_2,0) & (S_5,-0.41) & (S_4,0) & (S_4,0.22) \\ (S_4,0.19) & (S_5,0.03) & (S_4,-0.22) & (S_4,0) \end{pmatrix}$$

Step 3: By using the linguistic fuzzy quantifier "most" with the pair $(0.0, 0.5)$, the weight vector is produced as $\omega^{*}=(0.6478, 0.2355, 0.0856,\ 0.0311)^T$. From eq. (6.5), the degree of how much the decision alternative x_i is superior to another decision alternative x_j, $j\in N$, is established as follows:

$$t_1=(S_5,0.28),\ t_2=(S_4,0.07),\ t_3=(S_4,0.37),\ t_4=(S_5,-0.29),$$

Obviously, we have $t_1 > t_4 > t_3 > t_2$.

Step 4: The ranking of the four decision alternatives is $x_1 \succ x_4 \succ x_3 \succ x_2$.

6.4 Conclusions

Two-tuple linguistics can solve the problem of having the available information be distorted in the process of aggregating the information. Based on the principle that 2-tuple linguistic terms and the corresponding trapezoidal fuzzy numbers are equivalent, in order to overcome the uncertainty problem when trapezoidal fuzzy numbers are compared, we proposed in this chapter a priority approach to addressing preference relations of trapezoidal fuzzy numbers based on the 2-tuple linguistics. We discussed the transformation relationship between trapezoidal fuzzy numbers and 2-tuple linguistics, and established the relationship between these two. By using the relationship between preference relations of trapezoidal fuzzy numbers and 2-tuple LPRs, we developed a priority method of maximum entropy for group judgment information that takes the form of trapezoidal fuzzy preference. Then we provided the particular steps for aggregating collective preference relation of trapezoidal fuzzy numbers. At the end, we illustrated by a numerical example that our proposed method could effectively avoid the problem of information lose and distortion that occur during the process of aggregating the appraisal information of trapezoidal fuzzy numbers.

Chapter 7
Group Decision Making for Different Fuzzy Preference Relations

In each group decision making, due to the interactions of the different points of view or individual characters, even for the same decision problem, decision makers may present different judgments of preference with uncertainty values as entries, such as fuzzy numbers, triangular fuzzy numbers, interval fuzzy numbers, trapezoidal fuzzy numbers (Xiao, Fan, and Wang, 2002; Guo and Guo, 2005; Wang and Chuu, 2004; Jose, 2006; Zadrozny and Kacprzyk, 2006; Pasi and Yager, 2006; Gong and Liu, 2007a; Gong, Wu, and Cui, 2008). In consequence, resolving conflicts that arise from different preferences and finding a unanimously agreed upon preference as the collective opinion of the group is a very important issue in any environment involving uncertainty. At the present, the topic of group decision making with different preferences has attached great attention and interest from a wide spectrum of scholars.

In the previous chapters, we discussed the properties and related theories of fuzzy preference relations (FPRs), complementary preference relations of interval fuzzy numbers (IFNCPRs), complementary preference relations of triangular fuzzy numbers (TFNCPRs), preference relations of trapezoidal fuzzy numbers and two-tuple linguistic preference relations (LPRs). In this chapter, we introduce a transformation function to establish the relationship between these preference relations (PRs), and show in the forms of theorems that this transformation function is reasonable. In the latter of this chapter, we will utilize the max-min and LOWA operators to aggregate different kinds of PRs for the purpose of group decision making.

7.1 Group Decision Making Based on 2-Tuple Linguistic LPRs with Different Fuzzy Preferences

7.1.1 Definitions

For the sake of convenience of our communication, we restate the definitions of two-tuple LPRs, IFNCPRs, TFNCPRs, as initially detailed in the previous chapters.

Z. Gong et al.: Uncertain Fuzzy Preference Relations, STUDFUZZ 281, pp. 103–120.
springerlink.com © Springer-Verlag Berlin Heidelberg 2013

To this end, let $[x^-, y^+]$ and (a,b,c) be an interval and a triangular fuzzy number, respectively, satisfying $0 \le x^-, y^+, a,b,c \le 1$. Each real number m can be denoted as either $m=[m,m]$ or $m=(m,m,m)$ such that

$$[x^-, y^+] \ge m \Leftrightarrow x^- \ge m, y^+ \ge m; \ (a,b,c) \ge m \Leftrightarrow a \ge m, b \ge m, c \ge m.$$

Definition 7.1 (Wu, 2004). A preference relation $\overline{R} = (\overline{r}_{ij})_{n\times n}$ is referred to as a complementary preference relation of interval fuzzy numbers (IFNCPR),, if the following are satisfied:

$$\overline{r}_{ii} = [0.5, 0.5] \tag{7.1}$$

$$r_{ijl} + r_{jiu} = r_{iju} + r_{jil} = 1 \tag{7.2}$$

where $\overline{r}_{ij} = [r_{ijl}, r_{iju}], i, j \in N$, representing the range of membership degree to which the decision alternative x_i is preferred to another decision alternative x_j.

Definition 7.2 (Jiang and Fan, 2002a, 2002b; Xu, 2003; 2004a). A preference relation $\tilde{R} = (\tilde{r}_{ij})_{n\times n}$ is referred to as a complementary preference relation of triangular fuzzy numbers (TFNCPR), if the following are satisfied:

$$\tilde{r}_{ii} = 0.\tilde{5} \tag{7.3}$$

$$r_{ijl} + r_{jiu} = r_{ijm} + r_{jim} = r_{iju} + r_{jil} = 1 \tag{7.4}$$

where $\tilde{r}_{ij} = (r_{ijl}, r_{ijm}, r_{iju}), i, j \in \mathrm{N}$, denoting the range of membership degree to which the decision alternative x_i is preferred to the decision alternative x_j, $i, j \in N$.

Definition 7.3. An IFNPR $\overline{R} = (\overline{r}_{ij})_{n\times n}$ is referred to as having the property of satisfactory consistency, if for all $i, j, k \in N, i \ne j \ne k$, the following hold true:

(1) $\overline{r}_{ij} \ge 0.5, \overline{r}_{jk} \ge 0.5$ imply $\overline{r}_{ik} \ge 0.5$; and
(2) $\overline{r}_{ij} \le 0.5, \overline{r}_{jk} \le 0.5$ imply $\overline{r}_{ik} \le 0.5$.

Definition 7.4. A linguistic matrix $T=(s_{ij})_{n\times n}$ is referred to as a preference relation of 2-tuple linguistic terms (LPR), if the following are satisfied:

$$s_{ii}=(p_{ii},\alpha_{ii})=(S_{\frac{g}{2}},0)$$
$$\Delta(\Delta^{-1}(s_{ij})+\Delta^{-1}(s_{ji}))=(s_{g},0)$$

where $s_{ij}=(p_{ij},\alpha_{ij}),p_{ij}\in S,\alpha_{ij}\in[-0.5,0.5),i,j\in N$.

Definition 7.5. A 2-tuple LPR $T=(s_{ij})_{n\times n}$ is said to have the property of satisfactory consistency, if for all $i,j,k\in N,i\neq j\neq k$, the following hold true:

$$(p_{ij},\alpha_{ij})\geq(S_{\frac{g}{2}},0),(p_{jk},\alpha_{jk})\geq(S_{\frac{g}{2}},0)\Rightarrow(p_{ik},\alpha_{ik})\geq(S_{\frac{g}{2}},0);$$
$$(p_{ij},\alpha_{ij})\leq(S_{\frac{g}{2}},0),\ (p_{jk},\alpha_{jk})\leq(S_{\frac{g}{2}},0)\Rightarrow(p_{ik},\alpha_{ik})\leq(S_{\frac{g}{2}},0).$$

where $s_{ij}=(p_{ij},\alpha_{ij}),\ p_{ij}\in S,-0.5\leq\alpha_{ij}<0.5),i,j,k\in N,i\neq j\neq k$.

In this section, both interval and triangular fuzzy numbers will be jointly known as also uncertainty values, and IFNPRs and TFNCPRs uncertainty preference relations. The property of satisfactory consistency of all kinds of preference relations is one of the important properties that reflect the characteristics of consistent thinking in human judgments.

7.1.2 Relationship between Uncertainty Preference Relations and 2-Tuple LPRs

The transformation function between uncertainty values and two-tuple linguistic values is given as follows (this subsection is mainly based on (Gong and Liu, 2007a)):

Definition 7.6 (Herrera, Martinez, and Sanchez, 2005). Let I be a uncertainty value, $S_T=\{S_0,S_1,\cdots,S_g\}$ a set of linguistic terms. Then the function τ_{IS_T} transforms I into a fuzzy set in S_T, where

$$\tau_{IS_T}:[0,1]\longrightarrow F(S_T)$$
$$\tau_{IS_T}(I)=\{(S_k,\omega_k)/k\in\{0,\cdots,g\}\}\tag{7.5}$$

$$\omega_k=\max_y min\{\mu_I(y),\mu_{S_k}(y)\}\tag{7.6}$$

where $F(S_T)$ is the set of all fuzzy sets defined in S_T, and $\mu_I(.)$ and $\mu_{S_k}(.)$ the membership functions associated with the uncertainty value I and the terms in S_k, respectively.

Definition 7.7 (Herrera, Martinez, and Sanchez, 2005). Let $\tau(I)=\{(S_0,\omega_0),\cdots,(S_g,\omega_g)\}$ be a fuzzy set that represents an uncertainty value I over the set $S=\{S_0,S_1,\cdots,S_g\}$ of linguistic terms. Then an uncertainty value by means of the following function χ that represents the information of the fuzzy set $\tau(I)$ can be obtained, where

$$\chi: F(S_T)\longrightarrow[0,g]$$

$$\chi\tau(I)=\chi(F(S_T))=\chi\{(S_j,\omega_j),j=0,1,\cdots,g\}=\sum_{j=0}^{g}j\omega_j(\sum_{j=0}^{g}\omega_j)^{-1}=\beta \quad (7.7)$$

That is, we can transform β in eq. (7.7) into its representative 2-tuple linguistic value according to Definition 5.2.

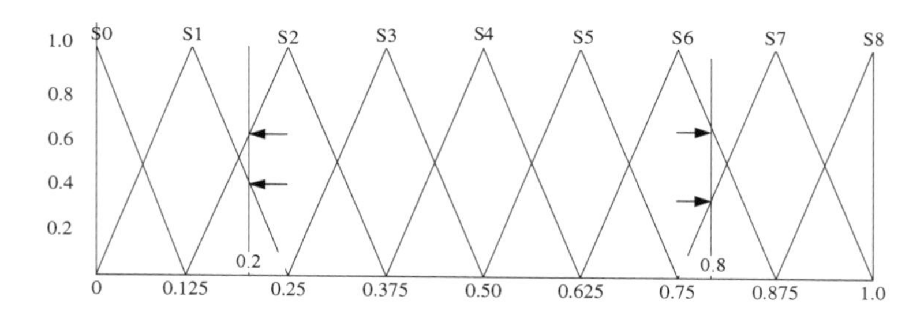

Fig. 7.1 Relationship between crisp fuzzy numbers and 2-tuple linguistic values

Example 7.1. Let $S=\{S_0,S_1,\cdots,S_8\}$ be a set of linguistic terms, see Figure 7.1 for related details, such that

$$\tau(0.2)=\{(S_0,0),(S_1,0.4),(S_2,0.6),(S_3,0),\cdots,(S_8,0)\},\chi\tau(0.2)=1\cdot0.4+2\cdot0.6=1.6,$$

Because $\Delta(1.6)=(S_2,-0.4)$, the transformation value of 0.2 is $(S_2,-0.4)$.

Similarly, the transformation value of 0.2 is $(S_6,0.4)$.

Thus we can conclude that when 0.2 and 0.8 are complementary (0.2+0.8=1), $\Delta(\Delta^{-1}(S_2,-0.4)+\Delta^{-1}(S_6,0.4))=(S_8,0)$.

In Figure 7.2, $[0.2,0.4]$ and $[0.6,0.8]$ are complementary, where

$$\tau[0.2,0.4] = \{(S_0,0),(S_1,0.4),(S_2,1),(S_3,1),(S_4,0.2),(S_5,0),\cdots,(S_8,0)\},$$
$$\chi\tau[0.2,0.4] = (1\cdot 0.4 + 2\cdot 1 + 3\cdot 1 + 4\cdot 0.2)(0.4+1+1+0.2)^{-1} = 2.38$$

So, from $\Delta(2.38) = (S_2,0.38)$, it follows that the transformation value of [0.2,0.4] is $(S_2,0.38)$.

Similarly, the transformation value of [0.6,0.8] is $(S_6,-0.38)$.

Therefore, we have $\Delta(\Delta^{-1}(S_2,0.38)+\Delta^{-1}((S_6,-0.38))) = (S_8,0)$.

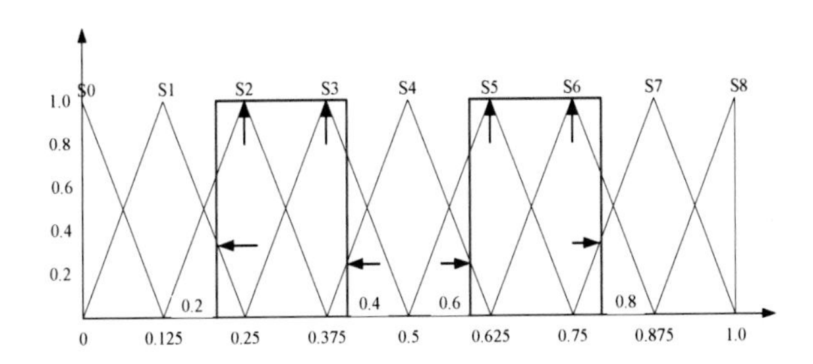

Fig. 7.2 Relationship between intervals and 2-tuple linguistic values

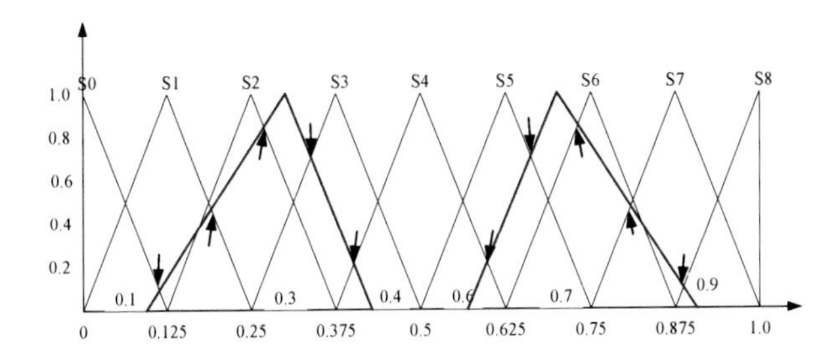

Fig. 7.3 Relationship between the triangular fuzzy numbers and 2-tuple linguistic values

In Figure 7.3, $(0.1,0.3,0.4)$ and $(0.6,0.7,0.9)$ are complementary, where

$$\tau(0.1,0.3,0.4) = \{(S_0,0.08),(S_1,0.46),(S_2,0.85),(S_3,0.67),(S_4,0.11),(S_5,0),\cdots,(S_8,0)\}$$
$$\chi\tau(0.1,0.3,0.4) = (1\cdot 0.46 + 2\cdot 0.85 + 3\cdot 0.67 + 4\cdot 0.11)(0.08+0.46+0.85+0.67+0.11)^{-1} = 2.13$$

From $\Delta(2.13) = (S_2,0.13)$, it follows that the transformation value of (0.1,0.3,0.4) is $(S_2,0.13)$.

Similarly, the transformation value of (0.6,0.7,0.9) is $(S_6, -0.13)$.

Therefore, we have $\Delta(\Delta^{-1}(S_2, 0.13) + \Delta^{-1}(S_6, -0.13)) = (S_8, 0)$.

Example 7.1 shows that if two values, no matter whether they are crisp numbers, intervals fuzzy or triangular fuzzy numbers, are complementary, then their transformed two-tuple linguistic values are also complementary. As is known, satisfactory consistency is one important property of all different kinds of preference relations.

In the following, we will show in the form of theorems that the transformed two-tuple LPRs of either IFNCPRs or TFNCPRs still have the property of satisfactory consistency (Gong and Liu, 2007a).

Theorem 7.1. The transformed 2-tuple LPR of an IFNPR in terms of the formulas in eqs. (7.5) - (7.7) still has the complementary property.

Proof. Let $S = \{S_0, \cdots, S_g\}$ be a set of linguistic terms, where

$$a_0 = 0, a_g = 1, a_{i+1} = b_i, b_{i+1} = c_i, \; i = 0, \cdots, g$$

and

$$a_i = \tfrac{i-1}{g}, \; b_i = \tfrac{i}{g}, \; c_i = \tfrac{i+1}{g}, \; c_i - b_i = b_i - a_i = \tfrac{1}{g}, i = 1, \cdots, g$$

Let $\overline{R} = (\overline{r}_{ij})_{n \times n}$ be an IFNPR satisfying Eqs. (7.1) and (7.2). Then, according to Eq.(7.5), we have

$$\tau(a_{ii}) = \tau[0.5, 0.5] = \tau(0.5) = \{(S_0, 0), (S_1, 0), \cdots, (S_{\frac{g}{2}-1}, 0), (S_{\frac{g}{2}}, 1), (S_{\frac{g}{2}+1}, 0), \cdots, (S_g, 0)\}$$

Thus, we have $\chi(\tau[0.5, 0.5]) = \frac{g}{2}$. That is, $\Delta(\frac{g}{2}) = (S_{\frac{g}{2}}, 0)$.

Let $r_{ijl} \in [b_i, c_i]$, $r_{iju} \in [a_j, b_j]$. Then we have $r_{jiu} \in [a_{g-i}, b_{g-i}]$ and $r_{jil} \in [b_{g-j}, c_{g-j}]$, where $j \geq i$. From Eq.(7.5), it follows that

$$\tau[r_{ijl}, r_{iju}] = \{(S_0, \omega_0), \cdots, (S_i, \omega_i), \cdots, (S_j, \omega_j), \cdots, (S_g, \omega_g)\}$$

where

$$\omega_0 = \cdots = \omega_{i-1} = \omega_{j+1} = \cdots = \omega_g = 0, \omega_{i+1} = \cdots = \omega_{j-1} = 1, \omega_i = g(c_i - r_{ijl}), \omega_j = g(r_{iju} - a_j)$$

$$\tau[a_{jil}, a_{jiu}] = \{(S_0, \omega_0'), \cdots, (S_{g-j}, \omega_{g-j}'), \cdots, (S_{g-i}, \omega_{g-i}'), \cdots, (S_g, \omega_g')\}$$

and

$$\omega_0' = \cdots = \omega_{g-j-1}' = \omega_{g-i+1}' = \cdots = \omega_g' = 0, \omega_{g-j+1}' = \cdots = \omega_{g-i-1}' = 1,$$
$$\omega_{g-j}' = g(c_{g-j} - r_{ji}^-),\ \omega_{g-i}' = g(r_{ji}^+ - a_{g-i})$$

Thus, we have

$$\beta_1 = \chi\tau[r_{ijl}, r_{iju}] = \sum_{j=0}^{g} j\omega_j (\sum_{j=0}^{g} \omega_j)^{-1}$$
$$= [ig(c_i - r_{ijl}) + (i+1) + \cdots + (j-1) + jg(r_{iju} - a_j)][g(c_i - r_{ijl}) + j - i - 1 + g(r_{iju} - a_j)]^{-1}$$
$$= [ig(c_i - r_{ijl}) + (i+1) + \cdots + (j-1) + jg(r_{iju} - a_j)][1 - g(r_{ijl} - r_{iju})]^{-1}$$
$$\beta_2 = \chi\tau[r_{jil}, r_{jiu}] = \sum_{j=0}^{g} j\omega_j' (\sum_{j=0}^{g} \omega_j')^{-1}$$
$$= [(g-j)g(c_{g-j} - r_{jil}) + (g-j+1) + \cdots + (g-i-1) + (g-i)g(r_{jiu} - a_{g-i})][1 - g(r_{ijl} - r_{iju})]^{-1}$$

It is ready to obtain that $\beta_1 + \beta_2 = g$, because $\Delta(g) = (S_g, 0)$. That is, $\Delta(\beta_1 + \beta_2) = (S_g, 0)$ and the transformed 2-tuple LPR has the complementary property. QED.

Lemma 7.1. If the interval $[a^-, a^+]$ satisfies $[a^-, a^+] \geq (\leq)[0.5, 0.5]$, then the transformed two-tuple linguistic value β of $[a^-, a^+]$ satisfies $\beta \geq (\leq)\frac{g}{2}$. That is, $\Delta(\beta) \geq (\leq)(S_{\frac{g}{2}}, 0)$.

Proof. Let $S = \{S_0, \cdots, S_g\}$ be a set of linguistic terms, where $a_0 = 0, a_g = 1, a_{i+1} = b_i, b_{i+1} = c_i$, $i = 0, \cdots, g$, such that

$$a_i = \frac{i-1}{g},\ b_i = \frac{i}{g},\ c_i = \frac{i+1}{g},\ c_i - b_i = b_i - a_i = \frac{1}{g}, i = 1, \cdots, g$$

Let $a^- \in [b_i, c_i]$, $a^+ \in [a_j, b_j]$, where $j \geq i$. Then according to the proof of Theorem 7.1, we have

$$\beta = \chi\tau[a^-, a^+] = \frac{\sum_{j=0}^{g} j\omega_j}{\sum_{j=0}^{g} \omega_j} = \frac{ig(c_i - a^-) + (i+1) + \cdots + (j-1) + jg(a^+ - a_j)}{1 - g(a^- - a^+)} = \frac{(i+j)\frac{i-j+1}{2} - iga^- + jga^+}{1 - g(a^- - a^+)}.$$

When $[a^-, a^+] \geq 0.5$, $\frac{g}{2} \leq i \leq g$, $\frac{g}{2} \leq j \leq g$, we have $\beta \geq \frac{g}{2}$. In the following, we will apply mathematical induction on the number $j - i = k$.

(1) When $j - i = 0$, we have $\beta_0 = [i + ig(a^+ - a^-)][1 + g(a^+ - a^-)]^{-1} \geq \frac{g}{2}$.

(2) Suppose that when $j - i = k$, we have $\beta_k \geq \frac{g}{2}$, that is,

$$\beta_k = [\tfrac{k+2i}{2}(1-k) - iga - +(k+i)ga^+][1 - g(a^- - a^+)]^{-1} \geq \tfrac{g}{2}.$$

(3) When $j - i = k + 1$, we have

$$\beta_{k+1} = [-k\tfrac{k+2i+1}{2} - iga^- + (k+i+1)ga^+][1 - g(a^- - a^+)]^{-1}$$
$$= \beta_k + \ (ga^+ - k - i)[1 - g(a^- - a^+)]^{-1}.$$

For $\frac{j-1}{g} \leq a^+ \leq \frac{j}{g}$, we have $ga^+ - k - i \geq j - 1 - k - i = 0$. Thus when $j - i = k + 1$, we have $\beta_{k+1} \geq \frac{g}{2}$. That is, when $[a^-, a^+] \geq 0.5$, $\beta \geq \frac{g}{2}$ holds.

Similarly, when $[a^-, a^+] \leq 0.5$, $\beta \leq \frac{g}{2}$ holds. QED.

In light of Lemma 7.1, we obtain the following result.

Theorem 7.2. The transformed 2-tuple LPR of an IFNPR in terms of Eqs. (7.5) - (7.7) has the property of satisfactory consistency.

Lemma 7.2. Let the set of linguistic terms be $S = \{S_0, \cdots, S_g\}$ that satisfies $S_i = (a_i, b_i, c_i)$, for $i = 0, \cdots, g$, and let $a_i \leq a \leq b_i, a_j \leq b \leq b_j, a_k \leq c \leq b_k$, $i \leq j \leq k$. Then, the following holds true:

$$\chi\tau(a,b,c) + \chi\tau(1-c, 1-b, 1-a) = g.$$

Proof. Because $a_i \leq a \leq b_i, a_j \leq b \leq b_j, a_k \leq c \leq b_k$, we have

$$b_{i-1} \leq a \leq c_{i-1}, b_{j-1} \leq b \leq c_{j-1}, b_{k-1} \leq c \leq c_{k-1}, i \leq j \leq k.$$

Based on the formula in eq. (7.5), let

$$\tau(a,b,c) = \{(S_0, \omega_0), \dots, (S_{i-2}, \omega_{i-2}), (S_{i-1}, \omega_{i-1}), \dots, (S_{j-1}, \omega_{j-1}), (S_j, \omega_j), \dots,$$
$$\dots, (S_k, \omega_k), (S_{k+1}, \omega_{K+1}), \dots, (S_g, \omega_g)\}$$

where $\omega_0 = \cdots = \omega_{i-2} = \omega_{k+1} = \cdots = \omega_g = 0$, $\omega_l = \frac{l+1-ga}{g(b-a)+1}$, $l = i-1, \cdots, j-1$, ω_l the ordinate of the intersection point of the line $y = \frac{x-a}{b-a}$ and the line $y = -g(x - \frac{l+1}{g})$; $\omega_m = \frac{-m+1+gc}{g(c-b)+1}, m = j, \cdots, k$, ω_m the ordinate of the intersection point of the line $y = -\frac{x-c}{c-b}$ and the line $y = g(x - \frac{m-1}{g})$.

Let

$$\tau(1-c,1-b,1-a) = \{(S_0, \omega_0'), \ldots, (S_{g-k-1}, \omega_{g-k-1}'), (S_{g-k}, \omega_{g-k}'), \ldots, (S_{g-j}, \omega_{g-j}'), (S_{g-j+1}, \omega_{g-j+1}'), \ldots,$$
$$\ldots, (S_{g-i+1}, \omega_{g-i+1}'), (S_{g-i+2}, \omega_{g-i+2}'), \ldots, (S_g, \omega_g')\}$$

where

$$\omega'_0 = \cdots, \omega'_{g-k-1} = \omega'_{g-i+2} = \cdots = \omega'_g = 0, \ \omega'_{l'} = \frac{l'+1-g(1-c)}{g(c-b)+1}, \ l' = g-k, \cdots, g-j,$$

$\omega'_{l'}$ is the ordinate of the intersection point of the line $y = \frac{x-1+c}{c-b}$ and the line $y = -g(x - \frac{l'+1}{g})$, $\omega'_{m'} = \frac{-m'+1+g(1-a)}{g(b-a)+1}, m' = g-j+1, \cdots, \ g-i+1$, $\omega_{m'}$ is the ordinate of the intersection point of the line $y = -\frac{x-1+a}{b-a}$ and the line $y = g(x - \frac{m'-1}{g})$.

Obviously, when $l = i-1, \cdots, j-1, m' = g-j+1, \cdots, \ g-i+1$, we have that $\omega_{i-1} = \omega'_{g-i+1}$, ..., $\omega_{j-1} = \omega'_{g-j+1}$; and when $m = j, \cdots, k, l' = g-k, \cdots, g-j$, we have $\omega_j = \omega'_{g-j}, \cdots, \omega_k = \omega'_{g-k}$. Consequently, we get

$$\begin{aligned}
&\chi\tau(a,b,c) + \chi\tau(1-c, 1-b, 1-a) \\
&= [\textstyle\sum_{l=i-1}^{j-1} \omega_l + \sum_{m=j}^{k} \omega_m]^{-1} \{[(i-1)\omega_{i-1} + \cdots + (j-1)\omega_{j-1}] + (j\omega_j + \cdots + k\omega_k) + \\
&\qquad [(g-k)\omega'_{g-k} + \cdots + (g-j)\omega'_{g-j}] + [(g-j+1)\omega'_{g-j+1} + \cdots + (g-i+1)\omega'_{g-i+1}]\} \\
&= [g(\textstyle\sum_{l=i-1}^{j-1} \omega_l + \sum_{m=j}^{k} \omega_m)][\sum_{l=i-1}^{j-1} \omega_l + \sum_{m=j}^{k} \omega_m]^{-1} = g.
\end{aligned}$$

Thus, we have $\Delta(g) = (S_g, 0)$, that is

$$\Delta(\chi\tau(a,b,c) + \chi\tau(1-c, 1-b, 1-a)) = (S_g, 0).$$ QED.

By employing Lemma 7.2, we can readily get the following result.

Theorem 7.3. The transformed 2-tuple LPR of a TFNPR in terms of the formulas in Eqs. (7.5) - (7.7) has the complementary property.

Lemma 7.3. If the triangular fuzzy number (a,b,c) satisfies $(a,b,c) \geq (\leq)(0.5,0.5,0.5)$, then the corresponding 2-tuple linguistic term as obtained by the transformation of Eqs. (7.5) - (7.7) satisfies $\beta \geq (\leq)\frac{g}{2}$ so that this 2-tuple linguistic term satisfies $\Delta(\chi\tau(a,b,c)) \geq (\leq)(S_{\frac{g}{2}},0)$.

The proof is similar to that of Lemma 7.2 and is omitted. QED.

In virtue of Lemma 7.3, we can readily conclude that

Theorem 7.4. The transformed 2-tuple LPR of a TFNPR in terms of the formulas in Eqs. (7.5) - (7.7) is still satisfactorily consistent. QED.

7.1.3 Aggregation of Different Preference Relations

In Chapter 5, we have studied the relationship between crisp fuzzy numbers and two-tuple linguistic values, while proving that the transformed two-tuple LPRs still have the property of additive consistency and the property of satisfactory consistency. According to Theorems 5.5 - 5.6, 6.1, and 7.1 - 7.4, when we aggregate the individual preference relations of the decision makers in the decision making group, we only need to transform these preference relations to two-tuple LPRs.

The specific procedure of aggregation is summarized as follows:

Step 1: Make the information contained in different preference relations uniform.

By utilizing formulas in eqs. (5.2), (7.5) - (7.7), transform the given preference relations $(f_{ij}^k)_{n\times n}$ of different kinds, such as FPRs, IFNCPRs, TFNCPRs, or preference relations of trapezoidal fuzzy numbers, into the 2-tuple LPRs $(S_{ij}^k,\alpha_{ij}^k)_{n\times n}$, $k=1,\cdots,m$; and transform the weights ω_k into the 2-tuple information $(f_k,\alpha_k), k=1,\cdots,m$.

Step 2: Aggregate the transformed 2-tuple LPRs.

Let the uniform preference relations be $P^k=(p_{ij}^k)_{n\times n}$, the weights of the decision makers be a_k, where $i,j=1,\cdots,n, k=1,\cdots,m$. Then the collective preference relation $P=(p_{ij})_{n\times n}$ of the group can be obtained by using the following formula:

$$P_{ij}=(S_{ij},\alpha_{ij})=\Delta(\frac{\sum_{k=1}^{m}\Delta^{-1}(S_{ij}^k,\alpha_{ij}^k).\Delta^{-1}(f_k,\alpha_k)}{\sum_{k=1}^{m}\Delta^{-1}(f_k,\alpha_k)})$$

Step 3: Calculate the degree of how much every decision alternative is superior to each of the other alternatives.

Let the degree of how much a decision alternative x_i is superior to another alternative x_j be t_i, $j = 1,\cdots,n$. Now, calculate t_i, $j = 1,\cdots,n$, by using *LOWA* operator (Yager,1992,2004; Herrera, Herrera-Viedma, and Verdegay, 1995; 1996a) as follows:

$$t_i = \phi_Q(p_{i1},\cdots,p_{in}) = \begin{cases} \Delta(\sum_{k=1}^{n} \omega_k \Delta^{-1}(p'_{ik})), & p_{ij} \text{ is a 2-tuple linguistic value} \\ \sum_{k=1}^{n} \omega_k p'_{ik}, & p_{ij} \text{ is a fuzzy number} \end{cases} \tag{7.8}$$

where p'_{ik} is the k th largest label in the set $\{p_{i1},\cdots,p_{in}\}$. The weight vector $\omega_k, k = 1,\cdots,q$ can be determined by: $\omega_k = Q(k/q) - Q(k-1/q)$, where $Q(r)$ stands for a fuzzy linguistic quantifier used to represent the fuzzy majority over dimension (Herrera, Herrera-Viedma, and Verdegay, 1995; 1996a) and is defined by:

$$Q(r) = \begin{cases} 0, & r < a, \\ (r-a)/(b-a), & a \le r \le b, \\ 1, & r > b. \end{cases}$$

Some typical linguistic fuzzy quantifiers include such terms as "most ", "at least half", "as many as possible ". And the parameters (a,b) are respectively (0.3,0.8), (0,0.5) and (0.5,1).

Step 4: Rank the decision alternatives.

7.1.4 A Numerical Example

Suppose that there are four site selections $X = \{x_1, x_2, x_3, x_4\}$ for an investment project, and that the set of linguistic terms is $S = \{S_0, S_1, \cdots, S_8\}$. Let the set of decision makers be $E = \{e_1, e_2, e_3, e_4\}$, and the weights of the decision makers be VL, VH, L and VL, respectively. Assume that the different preference relations as provided by the decision makers are given as follows:

$$\boldsymbol{P}_1=\begin{pmatrix} [0.5,0.5] & [0.6,0.9] & [0.5,0.7] & [0.7,0.9] \\ [0.1,0.4] & [0.5,0.5] & [0.6,0.8] & [0.2,0.4] \\ [0.3,0.5] & [0.2,0.4] & [0.5,0.5] & [0.2,0.5] \\ [0.1,0.3] & [0.6,0.8] & [0.5,0.8] & [0.5,0.5] \end{pmatrix}$$

$$\boldsymbol{P}_2=\begin{pmatrix} 0.5 & 0.7 & 0.6 & 0.8 \\ 0.3 & 0.5 & 0.4 & 0.6 \\ 0.4 & 0.6 & 0.5 & 0.7 \\ 0.2 & 0.4 & 0.3 & 0.5 \end{pmatrix}$$

$$\boldsymbol{P}_3=\begin{pmatrix} M & H & M & VH \\ L & M & L & H \\ M & H & M & VH \\ VL & L & VL & M \end{pmatrix}$$

$$\boldsymbol{P}_4=\begin{pmatrix} (0.5,0.5,0.5) & (0.6,0.7,0.8) & (0.5,0.6,0.7) & (0.6,0.8,0.9) \\ (0.2,0.3,0.4) & (0.5,0.5,0.5) & (0.1,0.3,0.4) & (0.2,0.4,0.5) \\ (0.3,0.4,0.5) & (0.6,0.7,0.9) & (0.5,0.5,0.5) & (0.1,0.2,0.3) \\ (0.1,0.2,0.4) & (0.5,0.6,0.8) & (0.7,0.8,0.9) & (0.5,0.5,0.5) \end{pmatrix}$$

Step 1: By utilizing the formulas in eqs. (5.2), (7.5) - (7.7), we obtain respectively the transformed 2-tuple linguistic values $(S_2,0),(S_6,0),(S_3,0),(S_2,0)$ of the decision makers weights, and the transformed 2-tuple LPRs of the individual preference relations P_1,P_2,P_3,P_4 as follows:

$$\boldsymbol{P}_1=\begin{pmatrix} (S_4,0) & (S_6,0) & (S_5,-0.15) & (S_6,0.38) \\ (S_2,0) & (S_4,0) & (S_6,-0.38) & (S_2,0.38) \\ (S_3,0.15) & (S_2,0.38) & (S_4,0) & (S_3,-0.24) \\ (S_2,-0.38) & (S_6,-0.38) & (S_5,0.24) & (S_4,0) \end{pmatrix}$$

$$\boldsymbol{P}_2=\begin{pmatrix} (S_4,0) & (S_6,-.4) & (S_5,-.2) & (S_6,.4) \\ (S_2,.4) & (S_4,0) & (S_3,.2) & (S_5,-.2) \\ (S_3,.2) & (S_5,-.2) & (S_4,0) & (S_6,-.4) \\ (S_2,-.4) & (S_3,.2) & (S_2,.4) & (S_4,0) \end{pmatrix}$$

$$P_3=\begin{pmatrix}(S_4,0) & (S_5,0) & (S_4,0) & (S_6,0)\\(S_3,0) & (S_4,0) & (S_3,0) & (S_5,0)\\(S_4,0) & (S_5,0) & (S_4,0) & (S_6,0)\\(S_2,0) & (S_3,0) & (S_2,0) & (S_4,0)\end{pmatrix}$$

$$P_4=\begin{pmatrix}(S_4,0) & (S_6,-0.375) & (S_5,-0.125) & (S_6,0.13)\\(S_2,0.375) & (S_4,0) & (S_2,0.13) & (S_3,-0.13)\\(S_3,0.125) & (S_6,-0.13) & (S_4,0) & (S_2,-0.375)\\(S_2,-0.13) & (S_5,0.13) & (S_6,0.375) & (S_4,0)\end{pmatrix}$$

Step 2: By utilizing the formula in eq. (5.4), we obtain the following collective linguistic preference relation:

$$P=\begin{pmatrix}(S_4,0) & (S_5,0.42) & (S_5,-0.5) & (S_6,0.20)\\(S_3,-0.42) & (S_4,0) & (S_3,0.3) & (S_4,0.30)\\(S_3,0.5) & (S_5,-0.3) & (S_4,0) & (S_5,-0.15)\\(S_2,-0.20) & (S_4,-0.30) & (S_3,0.15) & (S_4,0)\end{pmatrix}$$

Step 3: By using the linguistic fuzzy quantifier "most" on the pair $(0.3, 0.8)$, we obtain the weight vector $\omega = (0, 0.4, 0.5, 0.1)$. From eq. (7.8), it follows that the degree of how much the decision alternative x_i is superior to another decision alternative x_j, $j \in N$, is given as follows: $t_1 = (S_5, -0.18)$, $t_2 = (S_4, -0.49)$, $t_3 = (S_4, 0.23)$, $t_4 = (S_3, 0.23)$.

So, we have $t_1 > t_3 > t_2 > t_4$.

Step 4: The ranking of the four decision alternatives is $x_1 \succ x_3 \succ x_2 \succ x_4$.

7.2 Group Decision Making for Different Preference Relations

In this section, we discuss how different kinds of preference relations, such as TFNRPRs, the TFNCPRs, LPRs and the two-tuple LPRs, in group decision making scenarios can be aggregated.

7.2.1 Steps for Aggregating Different Fuzzy Preference Relations

The specific procedure of aggregation is summarized as follows, for more details, please consult with (Gong, Wu, and Cui, 2008):

Step 1: Uniform the information contained in different preference relations.

Step 1.1: Utilize the formula in eq. (4.21) to transform the given TFNRPRs into their corresponding TFNCPRs;

Step 1.2: Utilizing the formulas in eqs. (5.2), (7.5) and (7.7) to transform the given TFNCPRs into the corresponding 2-tuple LPRs; and

Step 1.3: Transform the weight information into the corresponding 2-tuple linguistic values (or triangular fuzzy number s).

Step 2: Aggregate the transformed preference relations.

Let the uniform preference relations be $P^k = (p_{ij}^k)_{n\times n}$, and the weights of the decision makers be a_k, where $i, j = 1, \cdots, n, k = 1, \cdots, m$. Then the collective preference relation $P = (p_{ij})_{n\times n}$ can be obtained by applying the following the max-min operator:

$$p_{ij} = \max_{k=1,\cdots,m} \min(p_{ij}^k, a_k) \tag{7.9}$$

Step 3: Calculate the degree of how much a decision alternative is superior to all other alternatives.

Let the degree of how much a decision alternative x_i is superior to another decision alternative x_j be t_i $j = 1, \cdots, n$. Then, apply *LOWA* (Herrera, Herrera-Viedma, and Verdegay, 1995; 1996a) operator to calculate t_i, $j = 1, \cdots, n$, as follows:

$$t_i = \phi_Q(p_{i1}, \cdots, p_{in}) = \begin{cases} \Delta(\sum_{k=1}^{n} \omega_k \Delta^{-1}(p_{ik}^{'})), & p_{ij} \text{ is a 2-tuple linguistic value} \\ \sum_{k=1}^{n} \omega_k p_{ik}^{'}, & p_{ij} \text{ is a the triangular fuzzy number} \end{cases} \tag{7.10}$$

where $p_{ik}^{'}$ is the k th largest label in the set $\{p_{i1}, \cdots, p_{in}\}$. The weight vector $\omega_k, k = 1, \cdots, q$, can be obtained from: $\omega_k = Q(k/q) - Q(k-1/q)$, where $Q(r)$ is a fuzzy linguistic quantifier used to represent the fuzzy majority over dimension (Yager,1992,2004; Herrera, Herrera-Viedma, and Verdegay, 1995; 1996a) and is determined by

$$Q(r) = \begin{cases} 0, & r < a, \\ (r-a)/(b-a), & a \le r \le b, \\ 1, & r > b. \end{cases}$$

Some typical linguistic fuzzy quantifiers include such terms as "most ", "at least half", "as many as possible ". The parameters (a,b) are respectively (0.3,0.8), (0.0.5), and (0.5,1).
Step 4: Rank the decision alternatives.

7.2.2 A Numerical Example

Consider a set $X = \{x_1, x_2, x_3, x_4\}$ of decision alternatives. Assume that the set of linguistic terms is $S = \{S_0, S_1, \cdots, S_8\}$, and the set of decision makers is $E = \{e_1, e_2, e_3, e_4\}$, and the weights of the individual decision makers are respectively VL,VH, L, and VL. Also assume that the individual preference relations as provided by the decision makers are as follows:

$$\boldsymbol{P}_1 = \begin{pmatrix} (1,1,1) & (\frac{3}{2},\frac{7}{3},4) & (1,\frac{3}{2},\frac{7}{3}) & (\frac{3}{2},4,9) \\ (\frac{1}{4},\frac{3}{7},\frac{2}{3}) & (1,1,1) & (\frac{1}{9},\frac{3}{7},\frac{2}{3}) & (\frac{1}{4},\frac{2}{3},1) \\ (\frac{3}{7},\frac{2}{3},1) & (\frac{3}{2},\frac{7}{3},9) & (1,1,1) & (\frac{1}{9},\frac{1}{4},\frac{3}{7}) \\ (\frac{1}{9},\frac{1}{4},\frac{2}{3}) & (1,\frac{3}{2},4) & (\frac{7}{3},4,9) & (1,1,1) \end{pmatrix}$$

$$\boldsymbol{P}_2 = \begin{pmatrix} (0.5,0.5,0.5) & (0.6,0.7,0.9) & (0.5,0.6,0.7) & (0.7,0.8,0.9) \\ (0.1,0.3,0.4) & (0.5,0.5,0.5) & (0.6,0.7,0.8) & (0.2,0.3,0.4) \\ (0.3,0.4,0.5) & (0.2,0.3,0.4) & (0.5,0.5,0.5) & (0.2,0.3,0.5) \\ (0.1,0.2,0.3) & (0.6,0.7,0.8) & (05,0.7,0.8) & (0.5,0.5,0.5) \end{pmatrix}$$

$$\boldsymbol{P}_3 = \begin{pmatrix} M & H & M & VH \\ L & M & L & H \\ M & H & M & VH \\ VL & L & VL & M \end{pmatrix}$$

$$\boldsymbol{P}_4 = \begin{pmatrix} (0.5,0.5,0.5) & (0.6,.7,.8) & (0.5,0.6,0.7) & (0.6,0.8,0.9) \\ (0.2,0.3,0.4) & (0.5,0.5,0.5) & (0.1,0.3,0.4) & (0.2,0.4,0.5) \\ (0.3,0.4,0.5) & (0.6,0.7,0.9) & (0.5,0.5,0.5) & (0.1,0.2,0.3) \\ (0.1,0.2,0.4) & (0.5,0.6,0.8) & (0.7,0.8,0.9) & (0.5,0.5,0.5) \end{pmatrix}$$

Approach 1: The transformed information is a two-tuple linguistic value

Step 1: Step 1.1: By utilizing the formula in eq. (4.21), we transform the matrix $\boldsymbol{P}_1$ to the following TFNCPR $\boldsymbol{P}_1^{'}$:

$$\boldsymbol{P}_1^{'}=\begin{pmatrix} (0.5,0.5,0.5) & (0.6,0.7,0.8) & (0.5,0.6,0.7) & (0.6,0.8,0.9) \\ (0.2,0.3,0.4) & (0.5,0.5,0.5) & (0.1,0.3,0.4) & (0.2,0.4,0.5) \\ (0.3,0.4,0.5) & (0.6,0.7,0.9) & (0.5,0.5,0.5) & (0.1,0.2,0.3) \\ (0.1,0.2,0.4) & (0.5,0.6,0.8) & (0.7,0.8,0.9) & (0.5,0.5,0.5) \end{pmatrix}$$

Step 1.2: By utilizing the formulas in eqs. (5.2), (7.5) - (7.7), we transform the matrices P_1, P_2, P_3 and P_4 to the following 2-tuple LPRs:

$$\boldsymbol{P}^{1}=\begin{pmatrix} (S_4,0) & (S_6,-0.375) & (S_5,-0.125) & (S_6,0.13) \\ (S_2,0.375) & (S_4,0) & (S_2,0.13) & (S_3,-0.13) \\ (S_3,0.125) & (S_6,-0.13) & (S_4,0) & (S_2,-0.375) \\ (S_2,-0.13) & (S_5,0.13) & (S_6,0.375) & (S_4,0) \end{pmatrix}$$

$$\boldsymbol{P}^{2}=\begin{pmatrix} (S_4,0) & (S_6,-0.13) & (S_5,-0.125) & (S_6,0.375) \\ (S_2,0.13) & (S_4,0) & (S_6,-0.375) & (S_2,-0.375) \\ (S_3,0.125) & (S_2,0.375) & (S_4,0) & (S_3,-0.39) \\ (S_2,-0.375) & (S_6,-0.375) & (S_5,0.39) & (S_4,0) \end{pmatrix}$$

$$\boldsymbol{P}^{3}=\begin{pmatrix} (S_4,0) & (S_5,0) & (S_4,0) & (S_6,0) \\ (S_3,0) & (S_4,0) & (S_3,0) & (S_5,0) \\ (S_4,0) & (S_5,0) & (S_4,0) & (S_6,0) \\ (S_2,0) & (S_3,0) & (S_2,0) & (S_4,0) \end{pmatrix}$$

$$\boldsymbol{P}^{4}=\begin{pmatrix} (S_4,0) & (S_6,-0.375) & (S_5,-0.125) & (S_6,0.13) \\ (S_2,0.375) & (S_4,0) & (S_2,0.13) & (S_3,-0.13) \\ (S_3,0.125) & (S_6,-0.13) & (S_4,0) & (S_2,-0.375) \\ (S_2,-0.13) & (S_5,0.13) & (S_6,0.375) & (S_4,0) \end{pmatrix}$$

Step 1.3: The transformed 2-tuple linguistic values of the decision makers weights are as follows: $(S_2,0),(S_6,0),(S_3,0),(S_2,0)$.

Step 2: By utilizing the formula in eq. (7.9), we obtain the following collective linguistic preference relation:

$$P = \begin{pmatrix} (S_4,0) & (S_6,-0.13) & (S_5,-0.125) & (S_6,0) \\ (S_3,0) & (S_4,0) & (S_6,-0.375) & (S_3,0) \\ (S_3,0.125) & (S_3,0) & (S_4,0) & (S_3,0) \\ (S_2,0) & (S_6,-0.375) & (S_5,0.39) & (S_4,0) \end{pmatrix}$$

Step 3: By using the linguistic fuzzy quantifier $Q(r)$ "most" on the pair $(0.3, 0.8)$, we obtain the weight vector $\omega = (0, 0.4, 0.5, 0.1)$. By employing eq. (7.10), we obtain the degree of how much a decision alternative x_i is superior to another alternative x_j, $j \in N$, as follows:

$$t_1 = (S_5, 0.19), t_2 = (S_3, 0.4), t_3 = (S_3, 0.05), t_4 = (S_4, 0.36).$$

which shows that $t_1 > t_4 > t_2 > t_3$.

Step 4: The ranking of the four decision alternatives is $x_1 \succ x_4 \succ x_2 \succ x_3$.

Approach 2: The transformed information is triangular fuzzy number

Step 1: Step 1.1: By utilizing the formula in eq. (4.21), we transform the matrices P_1, P_3 to the TFNCPRs $\boldsymbol{P}_1'$ and $\boldsymbol{P}^3$, and we let $\boldsymbol{P}^2 = \boldsymbol{P}_2$, $\boldsymbol{P}^4 = \boldsymbol{P}_4$, where

$$\boldsymbol{P}^3 = \begin{pmatrix} (0.375, 0.5, 0.625) & (0.5, 0.625, 0.75) & (0.375, 0.5, 0.625) & (0.625, 0.75, 0.875) \\ (0.25, 0.375, 0.5) & (0.375, 0.5, 0.625) & (0.25, 0.375, 0.5) & (0.5, 0.625, 0.75) \\ (0.375, 0.5, 0.625) & (0.5, 0.625, 0.75) & (0.375, 0.5, 0.625) & (0.625, 0.75, 0.875) \\ (0.125, 0.25, 0.375) & (0.25, 0.375, 0.5) & (0.125, 0.25, 0.375) & (0.375, 0.5, 0.625) \end{pmatrix}$$

Step 1.2: The transformed triangular fuzzy numbers of the decision makers weights are as follows: $(0.125, 0.25, 0.375)$, $(0.625, 0.75, 0.875)$, $(0.25, 0.375, 0.5)$, $(0.125, 0.25, 0.375)$.

Step 2: By utilizing the formula in eq. (7.9), we obtain the collective TFNCPR as follows:

$$P=\begin{pmatrix} (0.5,0.5,0.5) & (0.6,0.7,0.9) & (0.5,0.6,.7) & (0.625,0.75,0.875) \\ (0.25,0.375,0.5) & (0.5,0.5,0.5) & (0.6,0.7,0.8) & (0.25,0.375,0.5) \\ (0.3,0.4,0.5) & (0.25,0.375,0.5) & (0.5,0.5,0.5) & (0.25,0.375,0.5) \\ (0.125,0.25,0.375) & (0.6,0.7,0.8) & (0.5,0.7,0.8) & (0.5,0.5,0.5) \end{pmatrix}$$

Step 3: By using the linguistic fuzzy quantifier $Q(r)$ "most" on the pair $(0.3,0.8)$, we obtain the weight vector $\omega=(0,0.4,0.5,0.1)$. From eq. (7.10), it follows that the degree of how much a decision alternative x_i is superior to another alternative x_j, $j\in N$, is given below:

$t_1=(0.54,0.63,0.76)$, $t_2=(0.35,0.43,0.5)$, $t_3=(0.27,0.39,0.5)$, $t_4=(0.46,0.56,0.61)$.

which implies that $t_1>t_4>t_2>t_3$.

Step 4: The ranking of the four decision alternatives is $x_1\succ x_4\succ x_2\succ x_3$, which is the same what is obtained earlier.

7.3 Conclusions

Different kinds of preference relations must be unified or normalized in each group decision making. To this end, we can transform the available preference relations into either TFNCPRs or two-tuple LPRs. In this paper, we introduced a transformation function that can be employed to transform different kinds of preference relations into two-tuple LPRs. We then showed in the form of theorems that the transformation function is reasonable in the following sense: The transformed two-tuple LPRs from the original FPRs, TFNCPRs, or IFNCPRs still have the complementary property and the property of satisfactory consistency. Along with the theoretical investigation, we also established the detailed steps for aggregating different kinds of preference relations that most likely appear in group decision making scenarios.

Chapter 8
Intuitionistic Fuzzy Preference Relations

In any fuzzy preference relation (FPR), each element denotes the membership degree of how one decision alternative is preferred to another. The values of these elements range between 0 and 1. This key idea originates from Zadeh's concept of fuzzy sets (Zadeh, 1965). However, in many practical situations, it is hard for the decision makers to provide a precise judgment due to either the uncertainty of objective matters involved or the vague nature of human judgment. Hence many decision makers may prefer interval preference relations (IPRs) instead. And in any IPR, each element also denotes the range of the membership degree of how one decision alternative is preferred to another. These elements are characterized by using closed subintervals of $[0,1]$. This key idea comes from interval-valued fuzzy sets (IVFS) of Zadeh (Zadeh, 1975).

In 1986, Atanassov (Atanassov, 1986; Atanassov and Gargov,1989) generalized the concepts of Zadeh's fuzzy sets to intuitionistic fuzzy sets (IFS), where IFSs express the decision makers' subjective preferences by using membership degrees, non-membership degrees, and hesitation indexes. So, IFSs are well suited to dealing with inevitably imprecise or not totally reliable judgment, and have been applied in different research areas, such as medical diagnosis and decision-making problems (Bustince and Burillo, 1996; Dimitrov, 2004; Lei, Wang and Miao,2005; Li, 2005; Li, Wang and Liu, 2008; Lin, Yuan and Xia, 2007; Liu and Wang, 2007; Pankowska and Wygralak, 2006; Sugihara, Ishii and Tanaka, 2004; Supriya, Ranjit and Akhil, 2001; Szmidt and Kacprzyk, 2003; 2005a; 2005b; Tan and Zhang, 2006; Tizhoosh, 2008; Wang, 2009; Xu, 2007a; 2007b; Xu and Yager, 2006; Ye, 2009). Moreover, in the field of decision making serious attention has been given to the study of intuitionistic fuzzy set theory (Gong and Li, 2010; Gong, Li and Zhou, 2009; Gong, Li and Yao, 2010; Gong and Liu, 2007b; 2009; Tang and Gong, 2007; Yang and Chiclana).

As is known, the problems of arithmetic operations of intuitionistic fuzzy values have not been solved as of this writing. Even with these deficiencies, it did not prevent intuitionistic fuzzy sets to be utilized in the development of the decision making theory (Gong and Li, 2010; Gong, Li and Zhou, 2009; Gong, Li and Yao, 2010; Gong and Liu, 2007b; 2009; Hung and Yang, 2004; Li and Cheng, 2002; Li, Olson and Qin, 2007; Pankowska and Wygralak, 2006; Szmidt and Kacprzyk, 2000; 2004; Tang and Gong, 2007; Vlachos and Sergiadis, 2005;Wang, 2009; Yang and Chiclana,2009;Ye, 2009). Some scholars tried to avoid using the disputed

Z. Gong et al.: Uncertain Fuzzy Preference Relations, STUDFUZZ 281, pp. 121–193.
springerlink.com © Springer-Verlag Berlin Heidelberg 2013

arithmetic operations of intuitionistic fuzzy sets. in such cases, they used similarity measure, distance measure, and optimization measure to investigate problems of the decision making theory (Hung and Yang, 2004; Li and Cheng, 2002; Li, Olson and Qin, 2007; Liu and Wang, 2007; Pankowska and Wygralak, 2006; Sugihara, Ishii and Tanaka, 2004; Szmidt and Kacprzyk, 2000; 2004; Vlachos and Sergiadis, 2005; Wang, 2009; Yang and Chiclana, 2009; Ye, 2009), while others suggested to apply the constructed operation laws of intuitionistic fuzzy sets (Xu, 2007a; 2007b).

In this chapter, we investigate preference relations based on intuitionistic fuzzy sets from two different angles. On one hand, because each intuitionistic fuzzy set is equivalent to an interval-valued fuzzy set in terms of mathematical relations (Bustince and Burillo, 1996;Tizhoosh, 2008), we study intuitionistic fuzzy preference relations by transforming them into the equivalent interval fuzzy preference relations. This main content is contained throughout in Sections 8.1 - 8.8. On the other hand, we discuss intuitionistic fuzzy preference relations from the view point of similarity and correlation. This part of contents is developed in Sections 8.9 - 8.10.

8.1 Basic Concepts

In this section, after introducing the concept of intuitionistic fuzzy sets, we discuss an approach of comparing intuitionistic fuzzy numbers.

8.1.1 Intuitionistic Fuzzy Sets

Let $X=\{x_1,\dots,x_n\}$ be a finite non-empty set and $N=\{1,2,\dots,n\}$. A fuzzy set (Zadeh, 1965) F in X is an expression given by $F=\{<x,\mu_F(x)>| x\in X\}$, where $\mu_F : X\mapsto[0,1]$ is the membership function of F, and $\mu_F(x)\in[0,1]$ denotes the degree of membership of how much $x\in X$ in F.

An interval-valued fuzzy set (IVFS) (Zadeh, 1975) I in X is an expression written in the form of $I=\{<x,M_I(x)>| x\in X\}$, where $M_I : X\mapsto D[0,1]$ such that $M_I(x)=[M_{IL}(x),M_{IU}(x)]$, where $D[0,1]$ stands for the set of all closed subintervals of $[0,1]$, and $M_{IL}(x)$ and $M_{IU}(x)$ respectively the lower bound and upper bound of an interval of the interval $M_I(x)$. The concept of IVFSs is an extension of that of Zadeh's fuzzy sets.

Definition 8.1. (Intuitionistic fuzzy sets (Atanassov, 1986; Atanassov and Gargov, 1989)). Let $X=\{x_1,\dots,x_n\}$ be a finite non-empty set. An intuitionistic fuzzy set in X is an expression given by

$$A'=\{<x,\mu_{A'}(x),\nu_{A'}(x)>| x\in X\},$$

where $\mu_{A'}: X \mapsto [0,1]$ and $\nu_{A'}: X \mapsto [0,1]$ satisfy the condition that $0 \le \mu_{A'}(x) + \nu_{A'}(x) \le 1$, for all x in X. The numbers $\mu_{A'}(x)$ and $\nu_{A'}(x)$ denote respectively the membership degree and the non-membership degree of the element x in A'.

For each finite intuitionistic fuzzy set A' in X, $\pi_{A'}(x) = 1 - \mu_{A'}(x) - \nu_{A'}(x)$ is known as an intuitionistic fuzzy index of A'; it stands for a hesitation degree of whether x belongs to A' or not. It is obviously that $0 \le \pi_{A'}(x) \le 1$ holds true for each $x \in A$. If $\pi_{A'}(x) = 0$, then $\mu_{A'}(x) + \nu_{A'}(x) = 1$, which denotes that the intuitionistic fuzzy set A' is degenerated into the classic fuzzy set $A' = \{< x, \mu_{A'}(x) > | x \in X\}$ (Zadeh, 1965).

The concepts of IVFSs and IFSs are introduced on the bases of different semantics. However, from the mathematical point of view, the elements of an IVFS and the elements of an IFS can be transformed into each other (Bustinceand Burillo, 1996). Let $B' = \{< x, \mu_{B'}(x), \nu_{B'}(x) > | x \in X\}$ be an IFS, and $\pi_{B'}(x) = 1 - \mu_{B'}(x) - \nu_{B'}(x)$ an intuitionistic fuzzy index of B'. If we combine $\mu_{B'}(x)$ with $\pi_{B'}(x)$, and combine $\nu_{B'}(x)$ with $\pi_{B'}(x)$, then we can respectively produce the following two intervals:

$$B_1 = [\mu_{B'}(x), \mu_{B'}(x) + \pi_{B'}(x)] = [\mu_{B'}(x), 1 - \nu_{B'}(x)]$$

and

$$B_2 = [\nu_{B'}(x), \nu_{B'}(x) + \pi_{B'}(x)] = [\nu_{B'}(x), 1 - \mu_{B'}(x)].$$

Conversely, such intervals as $B_1 = [\mu_{B'}(x), 1 - \nu_{B'}(x)]$ and $B_2 = [\nu_{B'}(x), 1 - \mu_{B'}(x)]$ satisfying $\mu_{B'}(x) + \nu_{B'}(x) \le 1$ can be written as an IFS $B' = \{< x, \mu_{B'}(x), \nu_{B'}(x) > | x \in X\}$, as shown in Figure 8.1.

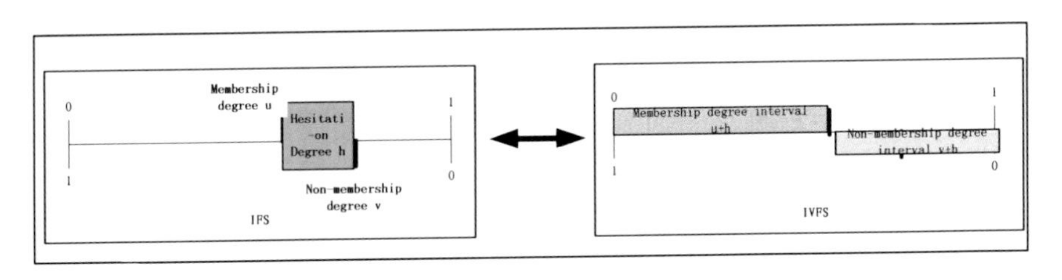

Fig. 8.1. Relationship between IFSs and IVFSs

8.1.2 Comparison of Intuitionistic Fuzzy Values

The so-called score of an intuitionistic fuzzy value $I = (\mu, \nu, \pi)$ is defined as $\Delta(I) = \mu - \nu$, where $\Delta(I) \in [-1,1]$; and the degree of accuracy of the intuitionistic fuzzy value is defined as $H(I)=\mu+\nu$, where $H(I)\in[0,1]$. The method of comparing two intuitionistic fuzzy values is given below (Xu, 2007a; 2007b).

Let $I_1=(\mu_1,\nu_1,\pi_1)$ and $I_2=(\mu_2,\nu_2,\pi_3)$ be two intuitionistic fuzzy values, $\Delta(I_1)=\mu_1-\nu_1$ and $\Delta(I_2)=\mu_2-\nu_2$ respectively the scores of I_1 and I_2. And let $H(I_1)=\mu_1+\nu_1$ and $H(I_2)=\mu_2+\nu_2$ be the accuracy degrees of I_1 and I_2, respectively. Then, we have

- If $\Delta(I_1)<\Delta(I_1)$, then I_1 is said to be smaller than I_2, denoted $I_1<I_2$; and
- If $\Delta(I_1)=\Delta(I_1)$, then
  1) If $H_1=H_2$, then I_1 and I_2 represent the same information, denoted $I_1=I_2$; and
  2) If $H_1<H_2$, then I_1 is said to be smaller than I_2, denoted $I_1<I_2$.

8.2 Intuitionistic Fuzzy Preference Relations

8.2.1 Intuitionistic Fuzzy Preference Relations

Assume for simplicity that $N=\{1,2,\ldots,n\}$ and $M=\{1,2,\ldots,m\}$. Let $S=\{s_1,\ldots,s_n\}$ be a finite set of decision alternatives and $E=\{e_1,\ldots,e_n\}$ the set of decision makers. The matrix $R=(r_{ij})_{n\times n}$ satisfying $r_{ii}=0.5$ and $r_{ij}+r_{ji}=1$ is known as a fuzzy preference relation (FPR) (Tanino, 1984), where r_{ij} is the degree of how much the decision alternative s_i is preferred to the alternative s_j. The equation $r_{ij}=0.5$ denotes that there is no difference between s_i and s_j, and $r_{ij}>0.5$ denotes that s_i is preferred to s_j. In many real world decision problems, the decision makers may prefer interval judgment matrices to precise judgments. Usually, an interval judgment matrix $R'=(r'_{ij})_{n\times n}$ satisfying $r'_{ii}=[0.5,0.5]$ and $r'_{ijl}+r'_{jiu}=r'_{iju}+r'_{jil}=1$ is also referred to as an interval fuzzy number complementary preference relation (IFNCPR), where $r'_{ij}=[r'_{ijl},r'_{iju}]$ denotes the range of degree of how much the decision alternative s_i is preferred to alternative s_j, and $r'_{ij}=[0.5,0.5]$ denotes indifference between the alternatives s_i and s_j, and $r'_{ij}>[0.5,0.5]$ denotes that the alternative s_i is preferred to s_j.

Definition 8.2. (Intuitionistic fuzzy preference relation (Dimitrov, 2004). An intuitionistic fuzzy preference relation in S can be defined as

$$R = \{< (s_i, s_j), \mu_R(s_i, s_j), \nu_R(s_i, s_j) > | (s_i, s_j) \in S \times S\},$$

where $\mu_R : S \times S \mapsto [0,1]$, $\nu_R : S \times S \mapsto [0,1]$, $\mu_R(s_i, s_j)$ stands for the degree of how much the decision alternative s_i is preferred to s_j, and $\nu_R(s_i, s_j)$ the degree of how much the decision alternative s_i is not preferred to s_j. Moreover, the inequality $0 \le \mu_R(s_i, s_j) + \nu_R(s_i, s_j) \le 1$ holds true for every $(s_i, s_j) \in S \times S$, $i, j \in N$.

Intuitionistic fuzzy preference relation can be represented by intuitionistic fuzzy judgment matrix.

Definition 8.3. (Intuitionistic fuzzy judgment matrix (IFPR) (Xu, 2007a; 2007b; Gong and Li, 2009; 2010)). Let R be an intuitionistic fuzzy preference relation in S. For any $i, j \in N$, let $\mu_{ij} = \mu_R(s_i, s_j)$ and $\nu_{ij} = \nu_R(s_i, s_j)$. If the following conditions hold,

$$r_{ii} = (0.5, 0.5, 0); \mu_{ij} = \nu_{ji}, \nu_{ij} = \mu_{ji}, \pi_{ij} = \pi_{ji}; \mu_{ij} + \nu_{ij} + \pi_{ij} = 1 \tag{8.1}$$

then

$$R = (u_{ij}, \nu_{ij}, \pi_{ij}) = \begin{pmatrix} (\mu_{11}, \nu_{11}, \pi_{11}) & (\mu_{12}, \nu_{12}, \pi_{12}) & \cdots & (\mu_{1n}, \nu_{1n}, \pi_{1n}) \\ (\mu_{21}, \nu_{21}, \pi_{21}) & (\mu_{22}, \nu_{22}, \pi_{22}) & \cdots & (\mu_{2n}, \nu_{2n}, \pi_{2n}) \\ \vdots & \vdots & \cdots & \vdots \\ (\mu_{n1}, \nu_{n1}, \pi_{n1}) & (\mu_{n2}, \nu_{n2}, \pi_{n2}) & \cdots & (\mu_{nn}, \nu_{nn}, \pi_{nn}) \end{pmatrix}$$

is referred to as an intuitionistic fuzzy judgment matrix. All the IFPRs referred to hereinafter are intuitionistic fuzzy judgment matrices.

For all $i, j \in N$, μ_{ij} is the degree of how much the decision alternative s_i is preferred to the alternative s_j, ν_{ij} the degree of how much s_i is not preferred to s_j, and the intuitionistic indices π_{ij} satisfy that the larger π_{ij} values are, the higher hesitation margin of how much s_i is preferred to s_j.

In the process of decision making, the decision maker can increase his evaluation by adding to the value of his intuitionistic index (Li, 2005). This means that his/her judgment actually lies in the closed intervals $[\mu_{ij}, \mu_{ij} + \pi_{ij}]$ and $[\nu_{ij}, \nu_{ij} + \pi_{ij}]$.

8.2.2 Relationship between IFPRs and the IFNCPRs

An IFPR can be split into three matrices as follows:

$$u=\begin{pmatrix}\mu_{11} & \mu_{12} & \cdots & \mu_{1n}\\ \mu_{21} & \mu_{22} & \cdots & \mu_{2n}\\ \vdots & \vdots & \cdots & \vdots\\ \mu_{n1} & \mu_{n2} & \cdots & \mu_{nn}\end{pmatrix};v=\begin{pmatrix}v_{11} & v_{12} & \cdots & v_{1n}\\ v_{21} & v_{22} & \cdots & v_{2n}\\ \vdots & \vdots & \cdots & \vdots\\ v_{n1} & v_{n2} & \cdots & v_{nn}\end{pmatrix};\pi=\begin{pmatrix}\pi_{11} & \pi_{12} & \cdots & \pi_{1n}\\ \pi_{21} & \pi_{22} & \cdots & \pi_{2n}\\ \vdots & \vdots & \cdots & \vdots\\ \pi_{n1} & \pi_{n2} & \cdots & \pi_{nn}\end{pmatrix}.$$

If u and π, and v and π are respectively combined, we derive the following two interval matrices:

$$A=(a_{ij})=(\mu_{ij},p_{ij})=\begin{pmatrix}[\mu_{11},p_{11}] & [\mu_{12},p_{12}] & \cdots & [\mu_{1n},p_{1n}]\\ [\mu_{21},p_{21}] & [\mu_{22},p_{22}] & \cdots & [\mu_{2n},p_{2n}]\\ \vdots & \vdots & \cdots & \vdots\\ [\mu_{n1},p_{n1}] & [\mu_{n2},p_{n2}] & \cdots & [\mu_{nn},p_{nn}]\end{pmatrix};$$

$$B=(b_{ij})=(v_{ij},q_{ij})=\begin{pmatrix}[v_{11},q_{11}] & [v_{12},q_{12}] & \cdots & [v_{1n},q_{1n}]\\ [v_{21},q_{21}] & [v_{22},q_{22}] & \cdots & [v_{2n},q_{2n}]\\ \vdots & \vdots & \cdots & \vdots\\ [v_{n1},q_{n1}] & [v_{n2},q_{n2}] & \cdots & [v_{nn},q_{nn}]\end{pmatrix},$$

where we denote

$$p_{ij}=1-v_{ij},q_{ij}=1-\mu_{ij} \tag{8.2}$$

By using Eqs. (8.2), we have that

$$[\mu_{ii},p_{ii}]=[0.5,0.5],\mu_{ij}+p_{ji}=p_{ij}+\mu_{ji}=1,i,j\in N, \tag{8.3}$$

$$[v_{ii},q_{ii}]=[0.5,0.5],v_{ij}+q_{ji}=q_{ij}+v_{ji}=1,i,j\in N, \tag{8.4}$$

which imply that both A and B are IFNCPRs. Both A and B can be regarded as the decomposed matrices of the given IFPR R. That is to say, the interval $[\mu_{ij},p_{ij}]$ can be regarded as the range of degree of how much the decision alternative s_i is preferred to the alternative s_j, and $[v_{ij},q_{ij}]$ the range of degree of how much the alternative s_i is not preferred to the alternative s_j.

Consider again the IFNCPRs A and B along with the conditions in Eqs. (8.3) and (8.4) holding true. Let $\pi_{ij}=p_{ij}-\mu_{ij}$. Then, Eqs. (8.2), (8.3) and (8.4) actually imply that

$$\mu_{ij}=v_{ji},v_{ij}=\mu_{ji},\quad p_{ij}=q_{ji},q_{ij}=p_{ji},\quad \pi_{ij}=\pi_{ji},\quad \mu_{ij}+v_{ij}+\pi_{ij}=1,i,j\in N \tag{8.5}$$

In consequence, the IFPR R can be considered as a combination of the IFNCPRs A and B.

This discussion leads to the following concept.

Definition 8.4 (The equivalent matrices)**.** The interval preference relations (IPRs) $A=(a_{ij})_{n\times n}=([\mu_{ij},p_{ij}])_{n\times n}$ and $B=(b_{ij})_{n\times n}=([\nu_{ij},q_{ij}])_{n\times n}$ satisfying the conditions in Eqs. (8.2), (8.3) and (8.4) are referred to as equivalent matrices of the given interval fuzzy preference relation (IFPR) R.

The matrices A and B are actually two complementary preference relations of interval fuzzy numbers (IFNCPRs). The discussion of R can be transformed to that of the equivalent matrices A and B.

8.3 Additive Consistency of IFPRs

The property of additively consistency of IFPRs items from that of IFNCPRs.

8.3.1 Additive Consistency of IFPRs

This subsection is mainly based on (Gong and Li, 2009). For more details, please consult with the original work and the references listed there.

Let $A=(a_{ij})_{n\times n}=([\mu_{ij},p_{ij}])_{n\times n}$ and $B=(b_{ij})_{n\times n}=([\nu_{ij},q_{ij}])_{n\times n}$ be the equivalent matrices of a given IFPR such that Eqs. (8.2), (8.3) and (8.4) hold true, for all $i,j\in N$. Suppose that both A and B are additively consistent IFNCPRs, then we have

$$\mu_{ij}+\mu_{jk}=\mu_{ik}+0.5 \tag{8.6}$$

$$p_{ij}+p_{jk}=p_{ik}+0.5 \tag{8.7}$$

$$\nu_{ij}+\nu_{jk}=\nu_{ik}+0.5 \tag{8.8}$$

and

$$q_{ij}+q_{jk}=q_{ik}+0.5 \tag{8.9}$$

In fact, based on Eq. (8.2), we can obtain that Eq. (8.6) holds true $\Leftrightarrow$ Eq. (8.9) holds true, and Eq. (8.7) holds true $\Leftrightarrow$ Eq. (8.8) holds true. In consequence, we have the following conclusion.

Theorem 8.1. The IFNCPR $A=(a_{ij})_{n\times n}$ is additively consistent $\Leftrightarrow$ the IFNCPR $B=(b_{ij})_{n\times n}$ is additively consistent.

If we subtract Eq. (8.6) from Eq. (8.7) or subtract Eq. (8.8) from Eq. (8.9), we can then obtain that

$$\pi_{ij} + \pi_{jk} = \pi_{ik} \tag{8.10}$$

By making use of Theorem 8.1 and Eq. (8.10), the concept of additively consistent IFPRs can be developed.

Definition 8.5. If a given IFPR R satisfies Eqs. (8.6) and (8.10), then R is said to have the property of additive consistency (transitivity). Those IFPRs that satisfy the property of additive consistency are referred to as additively consistent IFPRs.

Definition 8.6. Let $R = (r_{ij})_{n\times n} = (\mu_{ij}, \nu_{ij}, \pi_{ij})_{n\times n}$ be an IFPR, for all $i, j, k \in N$, $i \neq j \neq k$. If

- When $0.5 \leq \lambda < 1$, $\mu_{ij} \geq \lambda$ and $\mu_{jk} \geq \lambda$ imply $\mu_{ik} \geq \lambda$; and
- When $0 < \lambda \leq 0.5$, $\mu_{ij} \leq \lambda$ and $\mu_{jk} \leq \lambda$ imply $\mu_{ik} \leq \lambda$,

then *R* is said to have the property of general transitivity.

Theorem 8.2. Each additively consistent IFPR also has the property of general transitivity.

Proof. Let $R = (r_{ij})_{n\times n} = (\mu_{ij}, \nu_{ij}, \pi_{ij})_{n\times n}$ be an additively consistent IFPR. When $0.5 \leq \lambda < 1$, let us suppose that $\mu_{ij} \geq \lambda$, $\mu_{jk} \geq \lambda$. For $\mu_{ij} + \mu_{jk} = \mu_{ik} + 0.5$, we then have $\mu_{ik} \geq 2\lambda - 0.5 \geq \lambda$.

Similarly, when $0 < \lambda \leq 0.5$, if $\mu_{ij} \leq \lambda$,and $\mu_{jk} \leq \lambda$, we can then show $\mu_{ik} \leq \lambda$. QED.

In an IFPR, $\mu_{ij} \geq 0.5$ denotes that the decision alternative s_i is preferred or equivalent to the alternative s_j. If s_i is preferred (superior) or equivalent to s_j, and if s_j is preferred (superior) or equivalent to s_k, then the logical judgment of a decision maker should be that the alternative s_i is preferred (superior) or equivalent to s_k. In other word, if $\mu_{ij} \geq 0.5$ and $\mu_{jk} \geq 0.5$ holds true, we then must conclude $\mu_{ik} \geq 0.5$. This assumption is one of the least requirements of consistent judgment. Thus we have the following concept.

Definition 8.7. Let $R = (r_{ij})_{n\times n} = (\mu_{ij}, \nu_{ij}, \pi_{ij})_{n\times n}$ be an IFPR, for all $i, j, k \in N$, $i \neq j \neq k$. If

- The conditions $\mu_{ij} \geq (>)0.5$ and $\mu_{jk} \geq (>)0.5$ imply $\mu_{ik} \geq (>)0.5$; and
- The conditions $\mu_{ij} \leq (<)0.5$ and $\mu_{jk} \leq (<)0.5$ imply $\mu_{ik} \leq (<)0.5$,

then R is said to have the property of (restricted) weak transitivity.

The property of weak transitivity tells us that as a minimal logical requirement and a fundamental principle, an IFPR should at least possess the psychological characteristics of human beings. If we let $\lambda = 0.5$, then the following result follows immediately from Theorem 8.2.

Corollary 8.1. Each additively consistent IFPR also has the property of weak transitivity.

Let us consider a priority chain $s_{u_1} \succeq (\succ) s_{u_2} \succeq (\succ) \cdots \succeq (\succ) s_{u_n}$ in the set $S = \{s_1, s_2, \cdots, s_n\}$ of decision alternatives, where s_{u_i} denotes the ith alternative in the priority chain, and $s_{u_i} \succeq (\succ) s_{u_j}$ represents that the decision alternative s_{u_i} is preferred or equivalent (superior) to the alternative s_{u_j}. If a priority chain in $S = \{s_1, s_2, \cdots, s_n\}$ exists such that $s_{u_i} \succeq (\succ) s_{u_j}$ and $s_{u_j} \succeq (\succ) s_{u_k}$, then there must be $s_{u_i} \succeq (\succ) s_{u_j}$, for all $i, j, k \in N$, in other word, if the elements in IFPR satisfying $\mu_{ij} \geq 0.5$ and $\mu_{jk} \geq 0.5$, there must be $\mu_{ik} \geq 0.5$, for all $i, j, k \in N$, then the property of weak transitivity of IFPRs would be associated with a priority chain in the manner as stated in the following lemma.

Lemma 8.1. An IFPR has the property of (restricted) weak transitivity if and only if there exists a priority chain $s_{u_1} \succeq (\succ) s_{u_2} \succeq (\succ) \cdots \succeq (\succ) s_{u_n}$ in $S = \{s_1, s_2, \cdots, s_n\}$, where s_{u_i} denotes the ith decision alternative in the chain, and $s_{u_i} \succeq (\succ) s_{u_j}$ that the decision alternative s_{u_i} is preferred or equivalent to the alternative s_{u_j}.

According to Lemma 8.1, if there exists a circulation $s_{u_{i0}} \succeq (\succ) \cdots \succeq (\succ) \cdots \succeq (\succ) s_{u_{i0}}$, then the corresponding preference relation on the set $S = \{s_1, s_2, \cdots, s_n\}$ of decision alternatives does not have the property of transitivity property, and the IFPR is inconsistent.

In the following, three different approaches of judging whether or not an IFPR R has the property of weak transitivity are given. To this end, let us first introduce an indicator matrix of R.

Let $R = (r_{ij})_{n \times n} = (\mu_{ij}, v_{ij}, \pi_{ij})_{n \times n}$ be an IFPR. We define an indicator matrix $E = (e_{ij})_{n \times n}$ of R as follows:

$$e_{ij} = \begin{cases} 1 & \mu_{ij} > 0.5, \\ 0 & \text{otherwise.} \end{cases}$$

Also, we define a sub-matrix series E_i, $\mu_{ij} \geq 0.5$, where E_i is derived from E_{i-1} by deleting one 0 row vector and one corresponding column vector of E_{i-1}, where it is assumed that $E_1 = E$ and $E_n = (0)$.

Theorem 8.3. An IFPR R has the property of weak transitivity if and only if there is at least one 0 row vector in E_i, for all $i \in N$.

Proof. Necessity. Suppose that the IFPR $R = (r_{ij})_{n \times n}$ defined on the set $S = \{s_1, s_2, \cdots, s_n\}$ of decision alternatives has the property of weak transitivity. From Lemma 8.1, it follows that R is associated with a priority chain $s_{u_1} \succeq s_{u_2} \succeq \cdots \succeq s_{u_n}$, where s_{u_i} denotes the ith decision alternative in the priority chain.

Because s_{u_n} is the most inferior decision alternative, we have that $\mu_{u_n j} \geq 0.5$, for all $j \in N$, and so $e_{u_n j} = 0$. That is, the entries of the u_nth row in E are all 0. Deleting the u_nth row and the u_nth column from E, we obtain a sub-matrix E_2 of $E_1 = E$. At this time, the priority relation of the remaining decision alternatives have not been changed, thus $\mu_{u_{n-1}}$ is the most inferior decision alternative of the these remaining alternatives. Obviously, in E_2, the entries of the row represented by $s_{u_{n-1}}$ are all 0. Deleting the u_nth row and the u_nth column, the u_{n-1}th row and the u_{n-1}th column of E, we obtain submatrix E_3. Continuing to this procedure, we eventually produce an $(n-1)$th sub-matrix

$$E_{n-1} = \begin{pmatrix} 0 & 1 \\ 0 & 0 \end{pmatrix} or \begin{pmatrix} 0 & 0 \\ 1 & 0 \end{pmatrix} or \begin{pmatrix} 0 & 0 \\ 0 & 0 \end{pmatrix}.$$

In E_{n-1}, let the 0 row vector be represented by s_{u_2}. Then by deleting the 0 row vector and the corresponding column from E_{n-1}, we have $E_n = (0)$, thus the most superior decision alternative s_{u_1} is obtained.

Sufficiency. Let the entries of the u_nth row vector in $E_1 = E$ be 0. It is readily seen that the decision alternative s_{u_n} is the most inferior. Now by deleting the u_nth row and the u_nth column in E_1, we obtain a sub-matrix E_2 of E_1. Let the 0 row vector of E_2 be represented by $s_{u_{n-1}}$, then $s_{u_{n-1}}$ is superior to s_{u_n}. Continuing this procedure, we eventually obtain the most superior decision alternative s_{u_1}. Thus a priority chain $s_{u_1} \succeq s_{u_2} \succeq \cdots \succeq s_{u_n}$ is obtained, where s_{u_i} denotes the ith superior decision alternative in the set $S = \{s_1, s_2, \cdots, s_n\}$ of all decision alternatives. That is, R has the property of weak transitivity. QED.

As a matter of fact, from the proof of Theorem 8.3, E_i, $i \in N$, can also be regarded as an indicator matrix of the given preference relation on the set $\{s_{u_1}, s_{u_2}, \cdots, s_{u_{n-i+1}}\}$ of decision alternatives.

Now, the following result follows immediately from Theorem 8.3.

Corollary 8.2. Each IFPR R has the property of restricted weak transitivity if and only if there is one and only one 0 row vector in E_i, for all $i \in N$.

Actually, according to Theorem 8.3 and Corollary 8.2, a priority algorithm for each (restricted) weak transitivity of IFPR R can be given below.

Algorithm 1.

Step 1: Construct the indicator matrix E_1 of R.
Step 2: Let $i = 1$.
Step 3: Search for the 0 row vector in the sub-matrix E_i. If a 0 row exists, then the decision alternative that represents this row is denoted $s_{u_{n-i+1}}$, and go to step 4. Otherwise go to step 5.
Step 4: Delete the 0 row in E_i (if there are more than one such rows, then randomly select one 0 row) and the corresponding column, and set $i = i+1$. If $i = n$, then the decision alternative that represents this row is denoted s_{u_1} (That is, R has the property of weak transitivity). End. Otherwise, go to step 3.
Step 5: The preference relation R is inconsistent. End.

In the following, we will discuss the property of weak transitivity of IFPRs from the point of view of graph theory. For the sake of convenience, let us first look at some basic terminology of the theory of digraphs (Li, 1988; Yang, 1989; Horn and Johnson, 1990).

A *digraph* $M = (m_{ij})_{n\times n}$, denoted $G(M)$, is such a directed graph that has n nodes $v_1, v_1, \ldots, v_n$ such that there is a directed arc in $G(M)$ from v_i to v_j if and only if $m_{ij} \neq 0$. If $\overline{v_i v_j}$ is an arc from v_i to v_j, then v_i is said to be adjacent to v_j and v_j is adjacent from v_i. The *outdegree* of a node v is the number of nodes adjacent from v, and the *indegree* of the node v is the number of nodes adjacent to v. An *indegree sequence* is denoted as $v_1^{d_1} \cdots v_j^{d_j} \cdots v_n^{d_n}$, where $v_j^{d_j}$ denotes the indegree of the node v_j is d_j. A *directed path* ρ in a graph G is a sequence of arcs $v_{i_1}v_{i_2}, v_{i_2}v_{i_3}, v_{i_3}v_{i_4}, \ldots$ in G. The *ordered* list of nodes in the directed path ρ is $v_{i_1}, v_{i_2}, \ldots$. The *length* of a directed path is the number of successive arcs in the directed path. A *cycle* is a directed path that begins and ends at the same node.

Lemma 8.2 (Li, 1988; Yang, 1989; Horn and Johnson, 1990). Let v_i and v_j be two given nodes of a directed graph $G(M)$ of M. There exists a directed path of length l in $G(M)$ from v_i to v_j if and only if $(M^l)_{ij} \neq 0$, where $(M^l)_{ij}$ denotes the $(i,j)$the entry of $\boldsymbol{M}^l = \overbrace{M \times M \times \cdots \times M}^{l}$.

This lemma implies that there is only one longest directed path of length $n-1$ in $G(M)$ if and only if $\boldsymbol{M}^{n-1} \neq 0$ and $\boldsymbol{M}^n = 0$.

Theorem 8.4. An IFPR R has the property of restricted weak transitivity if and only if the indegree sequence of $G(E)$ is $v_{u_1}^0 \cdots v_{u_j}^{j-1} \cdots v_{u_n}^{n-1}$, where E is the indicator matrix of R, and $V = \{v_{u_1}, \cdots, v_{u_n}\}$ is the node set of $G(E)$.

Proof. Necessity. Suppose that the IFPR R defined on the set $S = \{s_1, s_2, \cdots, s_n\}$ of decision alternatives has the property of restricted weak transitivity. From Lemma 8.1, it follows that R is associated with a priority chain $s_{u_1} \succ s_{u_2} \succ \cdots \succ s_{u_n}$, which means that

$$s_{u_1} \succ s_{u_2}, s_{u_1} \succ s_{u_3}, \cdots, s_{u_1} \succ s_{u_n}; \cdots; s_{u_j} \succ s_{u_{j+1}}, s_{u_j} \succ s_{u_{j+2}}, \cdots, s_{u_j} \succ s_{u_n}; \cdots; s_{u_{n-1}} \succ s_{u_n}, \tag{8.11}$$

where s_{u_i} denotes the ith decision alternative in the priority chain. Now, Eq. (8.11) implies that the u_1u_2th, $\cdots$, u_1u_nth entries; $\cdots$; $u_ju_{j+1}th$, $\cdots$, u_ju_nth entries; $\cdots$; $u_{n-1}u_nth$ entries of the indicator matrix E of R are all equal to 1, and the remaining entries of E are 0. That shows that the arc set of $G(E)$ is $\{\overline{v_{u_1}v_{u_2}}, \cdots, \overline{v_{u_1}v_{u_n}}, \cdots, \overline{v_{u_j}v_{u_{j+1}}}, \cdots, \overline{v_{u_j}v_{u_n}}, \cdots, \overline{v_{u_{n-1}}v_{u_n}}\}$, as shown in Figure 8.2, where v_{u_j} denotes the node of $G(E)$ that is corresponding to the decision alternative s_{u_j}, for all $j \in N$. In consequence, the indegree sequence of $G(E)$ is $v_{u_1}^0 \cdots v_{u_j}^{j-1} \cdots v_{u_n}^{n-1}$.

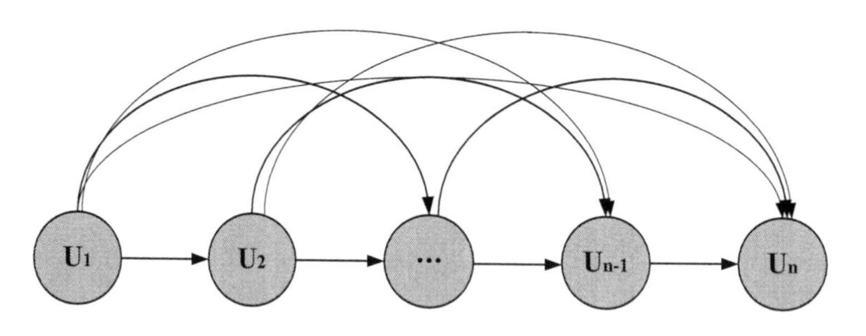

Fig. 8.2 Digraph $G(E)$

Sufficiency. The aforementioned process is reversible, so we omit the detailed sufficiency proof of this theorem. QED.

According to Theorem 8.4, a second priority algorithm for the property of restricted weak transitivity of a given IFPR R can be given as follows from a digraph point of view.

Algorithm 2.

Step 1: Construct the indicator matrix E of R.
Step 2: Construct the directed graph $G(E)$ of E.
Step 3: List the indegrees of all nodes in the ascending order: $v_{u_1}^{0} \cdots v_{u_j}^{j-1} \cdots v_{u_n}^{n-1}$.
Then the IFPR R has the property of restricted weak transitivity; otherwise, R is inconsistent.

Based on graph theory, a digraph with node-indegree series $n-1, n-2, \cdots, 2, 1$ does not have any cycle. Thus Theorem 8.4 can be rewritten as follows.

An IFPR R has the property of restricted weak transitivity if and only if there is no cycle in $G(E)$.

From Theorem 8.4, we can also conclude that there exists one directed path $v_{u_1} v_{u_2} \cdots v_{u_n}$ of length $n-1$ in digraph $G(E)$, and this is also the longest path in $G(E)$. The indicator matrix of the preference relation R on the set $\{s_{u_1} \cdots s_{u_n}\}$ of decision alternatives is E_1. By using Lemma 8.2, we can derive that $\boldsymbol{E}_1^{n-1} \neq 0$, $\boldsymbol{E}_1^{n} = 0$, and $(\boldsymbol{E}_1^{n-1})_{u_1 u_n} \neq 0$. If we delete the node v_{u_n}, then the length of the directed path $v_{u_1} \cdots v_{u_{n-1}}$ is $n-2$, and the indicator matrix of the preference relation on the set $\{s_{u_1} \cdots s_{u_{n-1}}\}$ of decision alternatives is E_2. By using Lemma 8.2, we can also derive that $\boldsymbol{E}_2^{n-2} \neq 0$, $\boldsymbol{E}_2^{n-1} = 0$ and $(E_2^{n-2})_{u_1 u_{n-1}} \neq 0$. Eventually, we obtain a directed path $v_{u_1} v_{u_2}$ of length 1, which implies that $\boldsymbol{E}_{n-1} \neq 0$, $\boldsymbol{E}_{n-1}^{2} = 0$, and $(\boldsymbol{E}_{n-1})_{u_1 u_2} \neq 0$. Obviously, this process is reversible. In consequence, we have derived the following result.

Theorem 8.5. An IFPR R has the property of restricted weak transitivity if and only if the indicator sub-matrix series E_i satisfies $\boldsymbol{E}_i^{n-i} \neq 0$, $\boldsymbol{E}_i^{n-i+1} = 0$, $(\boldsymbol{E}_i^{n-i})_{u_1 u_{n-i+1}} \neq 0, i = 1, 2, \ldots, n-1$.

On the basis of Theorem 8.5, we derive a third priority algorithm for weakly transitive IFPR R from the viewpoint of matrix analysis.

Algorithm 3.

The indicator sub-matrix series E_i can be constructed by using Algorithm 1. If there exists a matrix E_{i_0} such that $\boldsymbol{E}_{i_0}^{n-i_0+1} \neq 0$, then R is inconsistent. For all

$i=1,2,\ldots,n-1$, if $\boldsymbol{E}_i^{n-i}\neq 0$, $\boldsymbol{E}_i^{n-i+1}=0$, then R has the property of restricted weak transitivity.

According to Theorems 8.3, 8.4 and 8.5, there are three different ways to verify whether or not an IFPR has the property of restricted weak transitivity, of which the graph theory approach is the easiest and most straightforward. However, the digraph approach cannot express the possibility of indifference between two decision alternatives. This fact shows that the graph theory approach can only judge the property of restricted weak transitivity of a given IFPR R.

Definition 8.9. Let $R=(r_{ij})_{n\times n}=(\mu_{ij},\nu_{ij},\pi_{ij})_{n\times n}$ be an IFPR, for any $i,j,k\in N$, $i\neq j\neq k$. If $\mu_{ij}\geq 0.5$ and $\mu_{jk}\geq 0.5$ imply $\mu_{ik}\geq max\{\mu_{ij},\mu_{jk}\}$ and $p_{ik}\geq max\{p_{ij},p_{jk}\}$, then the IFPR R is said to have the property of restricted max-max transitivity.

Theorem 8.6. Each additively consistent IFPR R also has the property of restricted max-max transitivity.

Proof. Let $R=(r_{ij})_{n\times n}=(\mu_{ij},\nu_{ij},\pi_{ij})_{n\times n}$ be an additively consistent IFPR. Then for any $i,j,k\in N$, $i\neq j\neq k$, we have $\mu_{ij}+\mu_{jk}=\mu_{ik}+0.5, p_{ij}+p_{jk}=p_{ik}+0.5.$

If $\mu_{ij}\geq 0.5$ and $\mu_{jk}\geq 0.5$, then $p_{ij}\geq 0.5$ and $p_{jk}\geq 0.5$. Thus we can readily have

$$\mu_{ik}\geq\mu_{ij},\mu_{ik}\geq\mu_{jk};p_{ik}\geq p_{ij},p_{ik}\geq p_{jk}$$

That is, $\mu_{ik}\geq max\{\mu_{ij},\mu_{jk}\}$, $p_{ik}\geq max\{p_{ij},p_{jk}\}$. QED.

8.3.2 *Numerical Examples*

Example 8.1. Suppose that a decision maker is invited to assess four command-and-control systems $s_i, i=1,2,3,4$, on three readiness indexes, such as information accuracy, system availability, and picture completeness. The IFPR on the four alternatives is provided as follows:

$$R_1=\begin{pmatrix}(0.5,0.5,0.0) & (0.2,0.6,0.2) & (0.3,0.6,0.1) & (0.6,0.1,0.3)\\ (0.6,0.2,0.2) & (0.5,0.5,0.0) & (0.6,0.3,0.1) & (0.7,0.2,0.1)\\ (0.6,0.3,0.1) & (0.3,0.6,0.1) & (0.5,0.5,0.0) & (0.7,0.3,0.0)\\ (0.1,0.6,0.3) & (0.2,0.7,0.1) & (0.3,0.7,0.0) & (0.5,0.5,0.0)\end{pmatrix}.$$

Algorithm 1a.

Step 1: Construct the indicator matrix E_1 of R_1 as follows:

$$E_1 = \begin{pmatrix} 0 & 0 & 0 & 1 \\ 1 & 0 & 1 & 1 \\ 1 & 0 & 0 & 1 \\ 0 & 0 & 0 & 0 \end{pmatrix}.$$

Step 2: Let $E_{11} = E_1$.

Step 3: Search for a 0 row vector in E_{11}. The entries of the fourth row are all 0 so that s_4 is the most inferior alternative.

Step 4: By delete the fourth row and the fourth column in E_{11}, we get

$$E_{12} = \begin{pmatrix} 0 & 0 & 0 \\ 1 & 0 & 1 \\ 1 & 0 & 0 \end{pmatrix}.$$

Step 5: Search for a 0 row vector in E_{12}. The entries of the first row are al 0; and this row is also the first row of E_1. So s_1 is superior to s_4.

Step 6: By delete the fourth row and the fourth column, the first row and the first column in E_1, we get

$$E_{13} = \begin{pmatrix} 0 & 1 \\ 0 & 0 \end{pmatrix}.$$

Step 7: Search for a 0 row vector in E_{13}. The entries of the second row are all 0; and this row is the third row of E_1. So s_3 is superior to s_1.

Step 8: By delete the fourth row and the fourth column, the first row and the first column, the third row and the third column in E_1, we get E_{14}, where $E_{14} = (0)$.

Step 9: According to Theorem 8.3 and the corresponding algorithm, R_1 has the property of weak transitivity. So we conclude that s_2 is the most superior alternative.

Therefore, the priority chain of the set $\{s_1, s_2, s_3, s_4\}$ of decision alternatives is $s_2 \succ s_3 \succ s_1 \succ s_4$.

Algorithm 1b.

Step 1: Construct the indicator matrix E_1 of R_1 as follows:

$$E_1 = \begin{pmatrix} 0 & 0 & 0 & 1 \\ 1 & 0 & 1 & 1 \\ 1 & 0 & 0 & 1 \\ 0 & 0 & 0 & 0 \end{pmatrix}.$$

Step 2: Construct the digraph of $G(E_1)$ as shown in Figure 8.3.

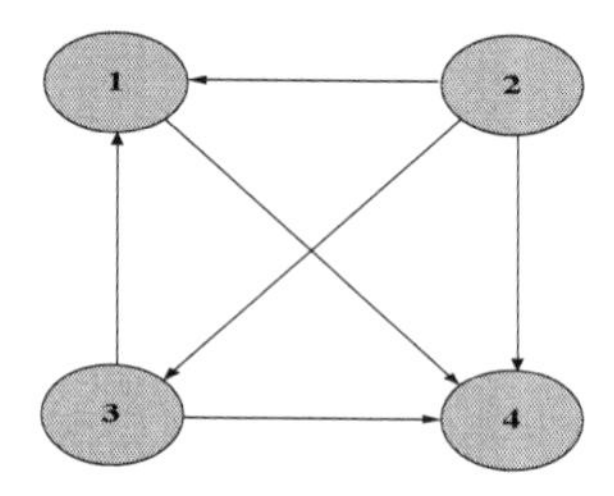

Fig. 8.3 Digraph $G(E_1)$

Step 3: Produce the indegree sequence of the digraph $G(E_1)$: $v_{s_1}^2 v_{s_2}^0 v_{s_3}^1 v_{s_4}^3$.

Step 4: From Theorem 4, it follows that R_1 has the property of restricted weak transitivity; and the priority chain of the set $\{s_1, s_2, s_3, s_4\}$ of decision alternatives is $s_2 \succ s_3 \succ s_1 \succ s_4$.

Algorithm 1c.

Step 1: By going through step 1 - step 6 of Algorithm 1a, we get the indicator matrix series E_{11}, E_{12}, E_{13}.

Step 2: From the facts that $E_{11}^3 \neq 0$, $E_{12}^2 \neq 0$, $E_{13} \neq 0$ and $E_{11}^4 = 0, E_{12}^3 = 0, E_{13}^2 = 0$, it follows that R_1 has the property of restricted weak transitivity.

Example 8.2. Suppose that a decision maker provides an IFPR R_2 on a set $S = \{s_1, s_2, s_3, s_4\}$ of decision alternatives as follows:

$$R_2 = \begin{pmatrix} (0.5,0.5,0.0) & (0.6,0.3,0.1) & (0.3,0.6,0.1) & (0.7,0.1,0.2) \\ (0.3,0.6,0.1) & (0.5,0.5,0.0) & (0.6,0.4,0.0) & (0.7,0.2,0.1) \\ (0.6,0.3,0.1) & (0.4,0.6,0.0) & (0.5,0.5,0.0) & (0.7,0.3,0.0) \\ (0.1,0.7,0.2) & (0.2,0.7,0.1) & (0.3,0.7,0.0) & (0.5,0.5,0.0) \end{pmatrix}.$$

Algorithm 2a.

Step 1: Construct the indicator matrix E_2 of R_2 as follows:

$$E_2 = \begin{pmatrix} 0 & 1 & 0 & 1 \\ 0 & 0 & 1 & 1 \\ 1 & 0 & 0 & 1 \\ 0 & 0 & 0 & 0 \end{pmatrix}.$$

Step 2: Let $E_{21} = E_2$.

Step 3: Search for a 0 row vector in E_{21}. The entries of the fourth row are all 0 so that s_4 is the most inferior alternative.

Step 4: By delete the fourth row and the fourth column in E_2, we get E_{22} as follows:

$$E_{22} = \begin{pmatrix} 0 & 1 & 0 \\ 0 & 0 & 1 \\ 1 & 0 & 0 \end{pmatrix}.$$

There is no longer any 0 row vector in E_{22}, which means that R_2 does not have the property of weak transitivity. In fact, in matrix E_2 we can see that s_1 is superior to s_2, s_2 is superior to s_3. So according to the requirements of consistent judgment, s_1 should be superior to s_3. However, in E_2 we can derive that s_3 is superior to s_1, which is a contradiction.

Algorithm 2b.

Step 1: Construct the indicator matrix E_2 of R_2 as given in Algorithm 2a.

Step 2: Construct the digraph $G(E_2)$ as shown in Figure 8.4. Obviously, there exists a cycle in $G(E_2)$. That shows that R_2 is inconsistent.

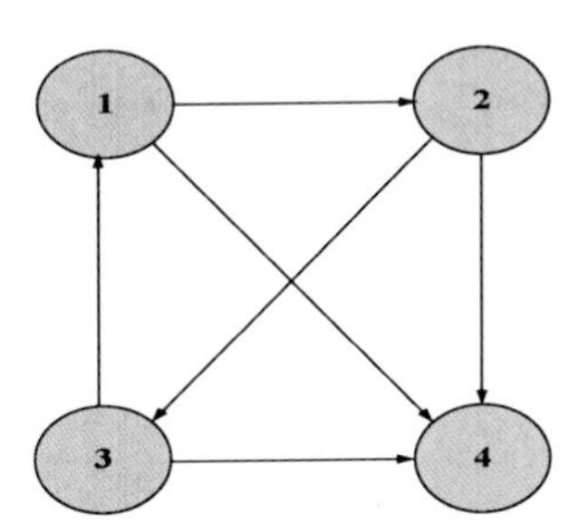

Fig. 8.4 Digraph $G(E_2)$

Algorithm 2c. Consider the indicator matrix E_2 of R_2. Because $\boldsymbol{E}_2^3 \neq 0; \boldsymbol{E}_2^4 \neq 0$, we know that R_2 is inconsistent.

From Examples 8.1 and 8.2, we can see that the graph theory approach is the easiest and the most straightforward one.

8.4 Multiplicatively Consistent IFPRs

Let $\omega=(\omega_1,\cdots,\omega_n)^T=([\omega_{1l}\ \omega_{1u}],\cdots,[\omega_{nl}\ \omega_{nu}])^T$ be the priority vector of the given consistent preference relation $A=(a_{ij})_{n\times n}=([\mu_{ij},p_{ij}])_{n\times n}$ of interval fuzzy numbers. Then for any $i,j\in N$, we have

$$a_{ij}=[\mu_{ij},p_{ij}]=\frac{1}{1+\frac{\omega_j}{\omega_i}}=[\frac{\omega_{il}}{\omega_{il}+\omega_{ju}},\frac{\omega_{iu}}{\omega_{jl}+\omega_{iu}}] \tag{8.12}$$

Because

$$\frac{1}{a_{ij}}-1=[\frac{1}{p_{ij}}-1,\frac{1}{\mu_{ij}}-1]=[\frac{\omega_{jl}}{\omega_{iu}},\frac{\omega_{ju}}{\omega_{il}}] \tag{8.13}$$

$$\frac{1}{a_{jk}}-1=[\frac{1}{p_{jk}}-1,\frac{1}{\mu_{jk}}-1]=[\frac{\omega_{kl}}{\omega_{ju}},\frac{\omega_{ku}}{\omega_{jl}}] \tag{8.14}$$

$$\frac{1}{a_{ki}}-1=[\frac{1}{p_{ki}}-1,\frac{1}{\mu_{ki}}-1]=[\frac{\omega_{il}}{\omega_{ku}},\frac{\omega_{iu}}{\omega_{kl}}] \tag{8.15}$$

we obtain

$$[\frac{1}{p_{ij}}-1][\frac{1}{p_{jk}}-1][\frac{1}{p_{ki}}-1]=\frac{\omega_{jl}}{\omega_{iu}}\frac{\omega_{kl}}{\omega_{ju}}\frac{\omega_{il}}{\omega_{ku}}=\frac{\omega_{il}}{\omega_{ju}}\frac{\omega_{jl}}{\omega_{ku}}\frac{\omega_{kl}}{\omega_{iu}}=[\frac{1}{p_{ji}}-1][\frac{1}{p_{kj}}-1][\frac{1}{p_{ik}}-1] \tag{8.16}$$

For the same reasons, we have

$$[\frac{1}{\mu_{ij}}-1][\frac{1}{\mu_{jk}}-1][\frac{1}{\mu_{ki}}-1]=[\frac{1}{\mu_{ji}}-1][\frac{1}{\mu_{kj}}-1][\frac{1}{\mu_{ik}}-1] \tag{8.17}$$

That is

$$[\frac{1}{a_{ij}}-1][\frac{1}{a_{jk}}-1][\frac{1}{a_{ki}}-1]=[\frac{1}{a_{ji}}-1][\frac{1}{a_{kj}}-1][\frac{1}{a_{ik}}-1] \tag{8.18}$$

For a given preference relation $B=(b_{ij})_{n\times n}=([\nu_{ij},q_{ij}])_{n\times n}$ of interval fuzzy numbers satisfying the conditions in Eqs. (8.3) and (8.5), combining Eqs. (8.6) with Eqs. (8.16) and (8.17) leads to

$$[\frac{1}{q_{ij}}-1][\frac{1}{q_{jk}}-1][\frac{1}{q_{ki}}-1]=[\frac{1}{q_{ji}}-1][\frac{1}{q_{kj}}-1][\frac{1}{q_{ik}}-1] \tag{8.19}$$

$$[\frac{1}{v_{ij}}-1][\frac{1}{v_{jk}}-1][\frac{1}{v_{ki}}-1]=[\frac{1}{v_{ji}}-1][\frac{1}{v_{kj}}-1][\frac{1}{v_{ik}}-1]=\frac{\omega_{iu}}{\omega_{jl}}\frac{\omega_{ju}}{\omega_{kl}}\frac{\omega_{ku}}{\omega_{il}} \tag{8.20}$$

That is, we have

$$[\frac{1}{b_{ij}}-1][\frac{1}{b_{jk}}-1][\frac{1}{b_{ki}}-1]=[\frac{1}{b_{ji}}-1][\frac{1}{b_{kj}}-1][\frac{1}{b_{ik}}-1] \tag{8.21}$$

Definition 8.10. The preference relations A or B of interval fuzzy numbers are said to have the property of multiplicative consistency, if, for any $i, j, k \in N$, Eq. (8.18) or (8.21) holds true. Any matrix satisfying the property of multiplicative consistency is referred to as multiplicative consistent.

The concept of multiplicatively consistent IFPRs is developed on the basis of Definition 8.10.

Definition 8.11. An intuitionistic fuzzy preference relation (IFPR) R is said to have the property of multiplicative consistency, if, for any $i, j, k \in N$, Eqs. (8.18) and (8.21) hold true. Any IFPR satisfying the property of multiplicative consistency is referred to as multiplicatively consistent.

In fact, by comparing each of the consistency conditions of these two preference relations of interval fuzzy numbers, we can conclude that Eq. (8.16) holds true $\Leftrightarrow$ Eq. (8.20) holds true; Eq. (8.17) holds true $\Leftrightarrow$ Eq. (8.19) holds true. That means that Eq. (8.18) holds $\Leftrightarrow$ Eq. (8.21) holds true. In consequence, we have derived the following conclusion.

Theorem 8.7. The matrix $A=(a_{ij})_{n\times n}=([\mu_{ij}, p_{ij}])_{n\times n}$ is multiplicatively consistent $\Leftrightarrow$ the matrix $B=(b_{ij})_{n\times n}=([v_{ij}, q_{ij}])_{n\times n}$ is multiplicatively consistent.

From Theorem 8.7, it follows that Definition 8.11 can be reformulated into a simpler expression as follows.

Definition 8.12. An IFPR R is said to have the property of multiplicative consistency, if, for any $i, j, k \in N$, Eq. (8.18) or (8.21) holds true.

As is well known, transitivity is an important issue with which many scholars have been concerned with. A consistent preference must satisfy the transitivity properties, such as restricted max-max transitivity, general transitivity, weak transitivity, etc. In the following, we will generalize the transitivity conditions of fuzzy preference relations to the case of IFPRs. Meanwhile, the relation between the property of multiplicative consistency and the transitivity properties of IFPRs is constructed to show that the defined consistent concept is reasonable and practically effective.

Definition 8.13. Let $A=(a_{ij})_{n\times n}=([\mu_{ij},p_{ij}])_{n\times n}$ be a preference relation of interval fuzzy numbers, for any $i,j,k\in N$, $i\neq j\neq k$. If $\mu_{ij}\geq\frac{1}{2}$ and $\mu_{jk}\geq\frac{1}{2}$,

- $\mu_{jk}\geq\mu_{ij}$ and $p_{jk}\geq p_{ij}$ imply either $\mu_{ik}\geq\mu_{jk}$ or $p_{ik}\geq p_{jk}$; and
- $\mu_{ij}\geq\mu_{jk}$ and $p_{ij}\geq p_{jk}$ imply either $\mu_{ik}\geq\mu_{ij}$ or $p_{ik}\geq p_{ij}$,

then A is said to have the property of restricted max-max transitivity.

From Eq. (8.5), it follows that Definition 8.13 actually implies that if $v_{ij}\leq\frac{1}{2}$ and $v_{jk}\leq\frac{1}{2}$, then

- $v_{jk}\leq v_{ij}$ and $q_{jk}\leq q_{ij}$ imply either $v_{ik}\leq\mu_{jk}$ or $q_{ik}\leq p_{jk}$; and
- $v_{ij}\leq v_{jk}$ and $q_{ij}\leq q_{jk}$ imply either $v_{ik}\leq\mu_{ij}$ or $q_{ik}\leq p_{ij}$.

The concept of restricted max-max transitive IFPRs is given below.

Definition 8.14. Let $R=(r_{ij})_{n\times n}=(\mu_{ij},v_{ij},\pi_{ij})_{n\times n}$ be an IFPR, for any $i,j,k\in N$, $i\neq j\neq k$.

- If $\mu_{ij}\geq\frac{1}{2}$ and $\mu_{jk}\geq\frac{1}{2}$, then
  - $\mu_{jk}\geq\mu_{ij}$ and $p_{jk}\geq p_{ij}$ imply either $\mu_{ik}\geq\mu_{jk}$ or $p_{ik}\geq p_{jk}$; and
  - $\mu_{ij}\geq\mu_{jk}$ and $p_{ij}\geq p_{jk}$ imply either $\mu_{ik}\geq\mu_{ij}$ or $p_{ik}\geq p_{ij}$,
- If $v_{ij}\geq\frac{1}{2}$ and $v_{jk}\geq\frac{1}{2}$, then
  - $v_{jk}\geq v_{ij}$ and $q_{jk}\geq q_{ij}$ imply either $v_{ik}\geq v_{jk}$ or $q_{ik}\geq p_{jk}$; and
  - $v_{ij}\geq v_{jk}$ and $q_{ij}\geq q_{jk}$ imply either $v_{ik}\geq v_{ij}$ or $q_{ik}\geq p_{ij}$,

then R is said to have the property of restricted max-max transitivity.

This definition implies that if the degree of how much the decision alternative s_i is preferred to the alternative s_j is the membership interval $a_{ij}=[\mu_{ij},p_{ij}]$, and the degree of how much the alternative s_j is preferred to a third alternative s_k is the membership interval $a_{jk}=[\mu_{jk},p_{jk}]$, then the degree of how much the alternative s_i is preferred to the alternative s_k is at least the lower limit of the membership interval $a_{ik}=[\mu_{ik},p_{ik}]$ or the upper limit of the membership interval $a_{ik}=[\mu_{ik},p_{ik}]$. If the degree of how much the alternative s_i is not preferred to the alternative s_j is the interval $b_{ij}=[v_{ij},q_{ij}]$, the degree of how much the alternative s_j is not preferred to the alternative s_k is the interval $b_{jk}=[v_{jk},q_{jk}]$, then the degree of how much the alternative s_i is not preferred to the alternative s_k is at least the lower limit of the interval $a_{ik} = [\mu_{ik}, p_{ik}]$ or the upper limit of the interval $a_{ik} = [\mu_{ik}, p_{ik}]$.

Lemma 8.3. Each multiplicatively consistent preference relation $A = (a_{ij})_{n\times n} = ([\mu_{ij}, p_{ij}])_{n\times n}$ of interval fuzzy numbers also has the property of restricted max-max transitivity.

Proof. For the case when $\mu_{ij} \geq \frac{1}{2}$, $\mu_{jk} \geq \frac{1}{2}$, we only need to prove that $\mu_{jk} \geq \mu_{ij}$ and $p_{jk} \geq p_{ij}$ imply either $\mu_{ik} \geq \mu_{jk}$ or $p_{ik} \geq p_{jk}$, for any $i, j, k \in N, i \neq j \neq k$.

Suppose for the purpose of producing a contradiction that there exist $i_0, j_0, k_0 \in N$, $i_0 \neq j_0 \neq k_0$, such that $\mu_{j_0k_0} \geq \mu_{i_0j_0}$ and $p_{j_0k_0} \geq p_{i_0j_0}$ imply $\mu_{i_0k_0} < \mu_{j_0k_0}$, $p_{i_0k_0} < p_{j_0k_0}$. So, from what is assumed, it follows that $1-\mu_{i_0k_0} > 1-\mu_{j_0k_0}$; that is, $p_{k_0i_0} > p_{k_0j_0}$. Therefore, we have

$$\frac{1}{p_{i_0k_0}} - 1 > \frac{1}{p_{j_0k_0}} - 1, \frac{1}{p_{k_0j_0}} - 1 > \frac{1}{p_{k_0i_0}} - 1 \tag{8.22}$$

Because $\mu_{i_0j_0} \geq \frac{1}{2}$, we have $p_{i_0j_0} \geq \frac{1}{2}$, $p_{j_0i_0} \leq \frac{1}{2}$, and thus

$$0 \leq \frac{1}{p_{i_0j_0}} - 1 \leq 1, \frac{1}{p_{j_0i_0}} - 1 \geq 1 \tag{8.23}$$

From Eqs. (8.16) and (8.23), it follows that

$$\begin{aligned}(\frac{1}{p_{j_0k_0}} - 1)(\frac{1}{p_{k_0i_0}} - 1) &\geq (\frac{1}{p_{i_0j_0}} - 1)(\frac{1}{p_{j_0k_0}} - 1)(\frac{1}{p_{k_0i_0}} - 1) \\ &= (\frac{1}{p_{j_0i_0}} - 1)(\frac{1}{p_{k_0j_0}} - 1)(\frac{1}{p_{i_0k_0}} - 1) \geq (\frac{1}{p_{k_0j_0}} - 1)(\frac{1}{p_{i_0k_0}} - 1)\end{aligned} \tag{8.24}$$

Meanwhile, from Eq. (8.22), we have

$$(\frac{1}{p_{j_0k_0}} - 1)(\frac{1}{p_{k_0i_0}} - 1) < (\frac{1}{p_{k_0j_0}} - 1)(\frac{1}{p_{i_0k_0}} - 1),$$

which contradicts Eq. (8.24). That is, condition $\mu_{ik} \geq \mu_{jk}$ or $p_{ik} \geq p_{jk}$ holds for any $i, j, k \in N, i \neq j \neq k$.

We can prove in a similar manner that when $\mu_{ij} \geq \mu_{jk}$ and $p_{ij} \geq p_{jk}$, we produce the conclusion of either $\mu_{ik} \geq \mu_{ij}$ or $p_{ik} \geq p_{ij}$. QED.

Lemma 8.3 means that for any $i, j, k \in N, i \neq j \neq k$, the upper limit of the membership degree satisfies $p_{ik} \geq max\{p_{ij}, p_{jk}\}$, the lower limit of the membership degree satisfies $\mu_{ik} \geq max\{\mu_{ij}, \mu_{jk}\}$.

The relationship between the property of multiplicative consistency and the restricted max-max transitivity of IFPRs can be established immediately from Lemma 8.3 as follows.

Theorem 8.8. Each multiplicatively consistent IFPR $R = (r_{ij})_{n\times n}$ also has the property of restricted max-max transitivity.

Definition 8.15. Let $A = (a_{ij})_{n\times n} = ([\mu_{ij}, p_{ij}])_{n\times n}$ be a preference relation of interval fuzzy numbers, for any $i, j, k \in N$, $i \neq j \neq k$. If

- when $0.5 \leq \lambda \leq 1$, the conditions $\mu_{ij} \geq \lambda$ and $\mu_{jk} \geq \lambda$ imply $p_{ik} \geq \lambda$;

and

- when $0 < \lambda \leq 0.5$, the conditions $p_{ij} \leq \lambda$ and $p_{jk} \leq \lambda$ imply $\mu_{ik} \leq \lambda$,

then A is said to have the property of general consistency.

Definition 8.15 implies that for any $i, j, k \in N, i \neq j \neq k$, when $0 < \lambda \leq 0.5$, the conditions $q_{ij} \leq \lambda$ and $q_{jk} \leq \lambda$ imply $v_{ik} \leq \lambda$, and when $0.5 \leq \lambda \leq 1$, the conditions $v_{ij} \geq \lambda$ and $v_{jk} \geq \lambda$ imply $q_{ik} \geq \lambda$.

Definition 8.16. Let $R = (r_{ij})_{n\times n}$ be an IFPR. If for any $i, j, k \in N, i \neq j \neq k$,

- when $0.5 \leq \lambda \leq 1$, the conditions $\mu_{ij} \geq \lambda$ and $\mu_{jk} \geq \lambda$ imply $p_{ik} \geq \lambda$; and the conditions $v_{ij} \geq \lambda$ and $v_{jk} \geq \lambda$ imply $q_{ik} \geq \lambda$; and
- when $0 < \lambda \leq 0.5$, the conditions $p_{ij} \leq \lambda$ and $p_{jk} \leq \lambda$ imply $\mu_{ik} \leq \lambda$; and the conditions $q_{ij} \leq \lambda$ and $q_{jk} \leq \lambda$ imply $v_{ik} \leq \lambda$,

then the IFPR R is said to have the property of general consistency.

What Definition 8.16 says is that if the degree of how much the decision alternative s_i is preferred to the alternative s_j satisfies the lower limit $\mu_{ij} \geq \lambda$, and the degree of how much the alternative s_j is preferred to the alternative s_k satisfies the lower limit $\mu_{jk} \geq \lambda$, then the degree of how much the alternative s_i is preferred to the alternative s_k satisfies at least that the upper limit $p_{ik} \geq \lambda$. If the degree of how much the alternative s_i is not preferred to the alternative s_j

satisfies the upper limit $p_{ij} \leq \lambda$, and the degree of how much the alternative s_j is not preferred to the alternative s_k satisfies the upper limit $p_{jk} \leq \lambda$, then the degree of how much the alternative s_i is not preferred to the alternative s_k satisfies at least that the lower limit $\mu_{ik} \leq \lambda$.

Theorem 8.9. Each multiplicatively consistent preference relation of interval fuzzy numbers has the property of general consistency.

Proof. Let $A = (a_{ij})_{n\times n} = ([\mu_{ij}, p_{ij}])_{n\times n}$ be a preference relation of interval fuzzy numbers with multiplicative consistency, for any $i, j, k \in N$, $i \neq j \neq k$. When $0.5 \leq \lambda \leq 1$, if $\mu_{ij} \geq \lambda$ and $\mu_{jk} \geq \lambda$ hold true, from Theorem 3.7, we obtain that $\mu_{ik} \geq \lambda$. When $0 < \lambda \leq 0.5$, if $p_{ij} \leq \lambda$ and $p_{jk} \leq \lambda$ hold true, then $p_{ji} \geq 1-\lambda$ and $p_{kj} \geq 1-\lambda$ follows. In the following, we verify $\mu_{ik} \leq \lambda$.

Suppose for the purpose of producing a contradiction that there exist $i_0, j_0, k_0 \in N$, $i_0 \neq j_0 \neq k_0$, such that when $0 < \lambda \leq 0.5$, the conditions $p_{i_0 j_0} \leq \lambda$ and $p_{j_0 k_0} \leq \lambda$ imply $\mu_{i_0 k_0} > \lambda$. Thus, we have $p_{i_0 k_0} > \lambda$, $\mu_{k_0 i_0} < 1-\lambda$, and $p_{k_0 i_0} < 1-\lambda$. Obviously, the following holds true:

$$\frac{1}{\mu_{i_0 j_0}} - 1 \geq \frac{1}{\lambda} - 1, \frac{1}{\mu_{j_0 k_0}} - 1 \geq \frac{1}{\lambda} - 1, \frac{1}{\mu_{j_0 i_0}} - 1 \leq \frac{1}{1-\lambda} - 1, \frac{1}{\mu_{k_0 j_0}} - 1 \leq \frac{1}{1-\lambda} - 1 \tag{8.25}$$

By using Eqs. (8.17) and (8.25), we have

$$\begin{aligned} \frac{\lambda}{1-\lambda}\frac{\lambda}{1-\lambda}\left(\frac{1}{\mu_{i_0 k_0}} - 1\right) &\geq \left(\frac{1}{\mu_{j_0 i_0}} - 1\right)\left(\frac{1}{\mu_{k_0 j_0}} - 1\right)\left(\frac{1}{\mu_{i_0 k_0}} - 1\right) \\ &= \left(\frac{1}{\mu_{i_0 j_0}} - 1\right)\left(\frac{1}{\mu_{j_0 k_0}} - 1\right)\left(\frac{1}{\mu_{k_0 i_0}} - 1\right) \geq \frac{1-\lambda}{\lambda}\frac{1-\lambda}{\lambda}\left(\frac{1}{\mu_{k_0 i_0}} - 1\right) \end{aligned} \tag{8.26}$$

That is, we have

$$\mu_{i_0 k_0} < \frac{\lambda^3}{\lambda^3 + (1-\lambda)^3}.$$

It is easy to see that when $0 < \lambda \leq 0.5$, the following inequality holds.

$$\frac{\lambda^3}{\lambda^3 + (1-\lambda)^3} \leq \lambda$$

That is, we have

$$\mu_{i_0k_0} < \frac{\lambda^3}{\lambda^3 + (1-\lambda)^3} \leq \lambda < \mu_{i_0k_0},$$

which is a contradiction. Therefore, when $0 < \lambda \leq 0.5$, if $p_{ij} \leq \lambda$ and $p_{jk} \leq \lambda$ hold true, we then have $\mu_{ik} \leq \lambda$. QED.

The relation between the property of multiplicative consistency and that of general consistency of IFPRs follows immediately from Theorem 8.9 as follows.

Corollary 8.3. Each multiplicative consistent IFPR R has also the property of general consistency.

Definition 8.17. Let $A = (a_{ij})_{n\times n} = ([\mu_{ij}, p_{ij}])_{n\times n}$ be an IFNCPR, for any $i, j, k \in N$, $i \neq j \neq k$. If

$$\mu_{ij} \geq \frac{1}{2}, \mu_{jk} \geq \frac{1}{2} \Rightarrow p_{ik} \geq \frac{1}{2},$$

then the IFNCPR A is said to have the property of weak consistency.

Similarly, we can establish the concept of weak consistency for an IFPR R.

Definition 8.18. Let $R = (r_{ij})_{n\times n}$ be an intuitionistic fuzzy preference relation (IFPR), for any $i, j, k \in N$, $i \neq j \neq k$. If

$$\text{either } \mu_{ij} \geq \frac{1}{2},\ \mu_{jk} \geq \frac{1}{2} \Rightarrow p_{ik} \geq \frac{1}{2} \text{ or } \nu_{ij} \geq \frac{1}{2}, \nu_{jk} \geq \frac{1}{2} \Rightarrow q_{ik} \geq \frac{1}{2},$$

then the IFPR R is said to have the property of weak consistency.

If we let $\lambda = \frac{1}{2}$, then following result follows immediately from Corollary 8.3.

Corollary 8.4. Each multiplicatively consistent IFPR R has the property of weak consistency.

8.5 Least Squared Priority Models of IFPRs

In this section, we discuss the problem of priorities of multiplicatively consistent IFPRs and non-multiplicatively consistent IFPRs.

8.5.1 Priority Model of Consistent IFPRs

Consider the equivalent matrices $A = (a_{ij})_{n\times n} = ([\mu_{ij}, p_{ij}])_{n\times n}$ and $B = (b_{ij})_{n\times n} = ([\nu_{ij}, q_{ij}])_{n\times n}$ of the given IFPR $R = (r_{ij})_{n\times n} = (\mu_{ij}, \nu_{ij}, \pi_{ij})_{n\times n}$, for which the conditions in Eqs. (8.2), (8.3), and (8.4) hold true. Let $\omega = (\omega_1 \cdots \omega_n)^T = ([\omega_{1l}\ \omega_{1u}] \cdots [\omega_{nl}\ \omega_{nu}])^T$ be the priority vector of the consistent preference relation A of interval fuzzy numbers, then we have

$$a_{ij} = [\mu_{ij}, p_{ij}] = \frac{1}{1+\frac{\omega_j}{\omega_i}} = [\frac{\omega_{il}}{\omega_{il}+\omega_{ju}}, \frac{\omega_{iu}}{\omega_{jl}+\omega_{iu}}] \tag{8.27}$$

That is, the following hold true:

$$\mu_{ij} = \frac{\omega_{il}}{\omega_{il}+\omega_{ju}}, i, j \in \mathrm{N} \tag{8.28}$$

$$p_{ij} = \frac{\omega_{iu}}{\omega_{jl}+\omega_{iu}}, i, j \in \mathrm{N} \tag{8.29}$$

If there is a vector $\omega = (\omega_1 \cdots \omega_n)^T = ([\omega_{1l}\ \omega_{1u}] \cdots [\omega_{nl}\ \omega_{nu}])^T$ such that Eq. (8.28) or (8.29) holds true, then A is said to be non-multiplicatively consistent.

From Eq. (8.2), we have

$$v_{ij} = 1 - p_{ij} = 1 - \frac{\omega_{iu}}{\omega_{jl}+\omega_{iu}} = \frac{\omega_{jl}}{\omega_{jl}+\omega_{iu}} \tag{8.30}$$

$$q_{ij} = 1 - \mu_{ij} = 1 - \frac{\omega_{il}}{\omega_{il}+\omega_{ju}} = \frac{\omega_{ju}}{\omega_{il}+\omega_{ju}} \tag{8.31}$$

Thus, the following follows:

$$b_{ij} = [v_{ij}, q_{ij}] = \frac{1}{1+\frac{\omega_i}{\omega_j}} = [\frac{\omega_{jl}}{\omega_{iu}+\omega_{jl}}, \frac{\omega_{ju}}{\omega_{il}+\omega_{ju}}] \tag{8.32}$$

This means that $\omega = (\omega_1, \cdots, \omega_n)^T = ([\omega_{1l}\ \omega_{1u}], \cdots, [\omega_{nl}\ \omega_{nu}])^T$ is also the priority vector of $\boldsymbol{B} = (b_{ij})_{n \times n} = ([v_{ij}, q_{ij}])_{n \times n}$.

In consequence, we have shown the following result.

Theorem 8.10. If the IFNCPRs $A = (a_{ij})_{n \times n} = ([\mu_{ij}, p_{ij}])_{n \times n}$ and $B = (b_{ij})_{n \times n} = ([v_{ij}, q_{ij}])_{n \times n}$ satisfy Eqs. (8.2), (8.3), and (8.4), then both A and B have the same priority vector.

According to Eqs. (8.28) and (8.29), or Eqs. (8.30) and (8.31), we have

$$\pi_{ij} = p_{ij} - \mu_{ij} = \frac{\omega_{iu}}{\omega_{jl} + \omega_{iu}} - \frac{\omega_{il}}{\omega_{il} + \omega_{ju}}, i, j \in \mathrm{N} \tag{8.33}$$

Jointly Eqs. (8.29), (8.30), and (8.32) show that v_{ij} and $\pi_{ij}, i, j \in \mathrm{N}$, can be denoted by the priority vector $\boldsymbol{\omega} = (\omega_1 \ldots \omega_n)^T$ of $\boldsymbol{A}$. If we let $\boldsymbol{\omega} = (\omega_1 \ldots \omega_n)^T$ be the priority vector of the multiplicatively consistent IFNCPR $A = (a_{ij})_{n\times n} = ([\mu_{ij}, p_{ij}])_{n\times n}$, then the membership degree μ_{ij}, the non-membership degree v_{ij}, and the intuitionistic fuzzy index π_{ij} of R can be derived by using Eqs. (8.28), (8.30), and (8.33), respectively. In fact, we can transform the interval vector Ω into the form of the intuitionistic fuzzy numbers as follows:

$$\zeta = ((\omega_{1l}, 1-\omega_{1u}, \omega_{1u} - \omega_{1l}) \ldots (\omega_{il}, 1-\omega_{iu}, \omega_{iu} - \omega_{il}) \ldots (\omega_{nl}, 1-\omega_{nu}, \omega_{nu} - \omega_{nl}))^T$$

where ω_{il} is the membership degree of the importance of the decision alternative x_i, $1-\omega_{iu}$ the non-membership degree of the importance of the alternative x_i, and $\omega_{iu} - \omega_{il}$ the hesitation degree of the importance of the alternative $x_i, i \in N$.

Based on the discussion above, we can establish the following important concept.

Definition 8.19. If there exists a vector $\zeta = (\zeta_1 \cdots \zeta_n)^T$ such that Eqs. (8.28), (8.30), and (8.33) hold true, then $R = (r_{ij})_{n\times n} = (\mu_{ij}, v_{ij}, \pi_{ij})_{n\times n}$ is said to be a multiplicatively consistent IFPR, where $\zeta_i = (\omega_{il}, 1-\omega_{iu}, \omega_{iu} - \omega_{il}), i \in N$, and the vector ζ the priority vector of the multiplicatively consistent IFPR *R*, and what are in Eqs. (8.28), (8.30), and (8.33) the multiplicatively consistency conditions of the IFPR *R*.

This definition shows that the priority vector of an IFPR R can be obtained from its equivalent interval matrices. Because Eq. (8.33) is derived from Eqs. (8.28) and (8.28), and Eq. (8.30) from Eq. (8.29), Eqs. (8.28) and (8.29) are referred to as the multiplicative consistency conditions of the IFPR R. If there is no such vector $\zeta = (\zeta_1 \cdots \zeta_n)^T$ that Eqs. (8.28) and (8.29) hold true, then *R* is said to be not multiplicatively consistent.

Theorem 8.11. Let $\boldsymbol{\omega} = (\omega_1\ \omega_2 \cdots \omega_n)^T$ be the priority vector of the given multiplicatively consistent IFNCPRs $\boldsymbol{A} = (a_{ij})_{n\times n} = ([\mu_{ij}, p_{ij}])_{n\times n}$ and $\boldsymbol{B} = (b_{ij})_{n\times n} = ([v_{ij}, q_{ij}])_{n\times n}$, and Eqs. (8.2), (8.3), and (8.4) hold true for all $i, j \in N$. Then the priority vector of the multiplicatively consistent IFPR $\boldsymbol{R} = (r_{ij})_{n\times n} = (\mu_{ij}, v_{ij}, \pi_{ij})_{n\times n}$ is given by

$$\zeta = ((\omega_{1l}, 1-\omega_{1u}, \omega_{1u} - \omega_{1l}) \ldots (\omega_{il}, 1-\omega_{iu}, \omega_{iu} - \omega_{il}) \ldots (\omega_{nl}, 1-\omega_{nu}, \omega_{nu} - \omega_{nl}))^T$$

Proof. This result can be established by employing the relationship among $\mu_{ij}, v_{ij}, p_{ij}, \pi_{ij}$. QED.

8.5.2 Priority Model of Multiplicatively Consistent IFPRs

Let $\boldsymbol{A} = (a_{ij})_{n\times n} = ([\mu_{ij}, p_{ij}])_{n\times n}$ be the equivalent matrix of the IFPR $\boldsymbol{R} = (r_{ij})_{n\times n} = (\mu_{ij}, v_{ij}, \pi_{ij})_{n\times n}$, and $\boldsymbol{\omega} = (\omega_1\ \omega_2\ \cdots\ \omega_n)^T$ the priority vector of the consistent IFNCPR $A = (a_{ij})_{n\times n} = ([\mu_{ij}, p_{ij}])_{n\times n}$, for any $i, j, k \in N, i \neq j \neq k$. Then, we have

$$\frac{1}{\mu_{ij}} - 1 = \frac{\omega_{ju}}{\omega_{il}}, \frac{1}{p_{ij}} - 1 = \frac{\omega_{jl}}{\omega_{iu}}$$

For simplicity, we let

$$x_{ij} = \frac{1}{\mu_{ij}} - 1,\ y_{ij} = \frac{1}{p_{ij}} - 1.$$

That is, we have

$$\omega_{il} x_{ij} - \omega_{ju} = 0, \omega_{iu} y_{ij} - \omega_{jl} = 0 \tag{8.34}$$

For all $i, j, k \in \mathrm{N}$, $\omega_i = (\omega_{il}, \omega_{iu}), \omega_j = (\omega_{jl}, \omega_{ju}), \omega_k = (\omega_{kl}, \omega_{ku})$ are solutions to the following equations:

$$\boldsymbol{A}\begin{pmatrix}\omega_{il} & \omega_{iu} & \omega_{jl} & \omega_{ju} & \omega_{kl} & \omega_{ku}\end{pmatrix}^T = \begin{pmatrix} x_{ij} & 0 & 0 & -1 & 0 & 0 \\ 0 & y_{ij} & -1 & 0 & 0 & 0 \\ 0 & 0 & x_{jk} & 0 & 0 & -1 \\ 0 & 0 & 0 & y_{jk} & -1 & 0 \\ 0 & -1 & 0 & 0 & x_{ji} & 0 \\ -1 & 0 & 0 & 0 & 0 & y_{ji} \end{pmatrix}\begin{pmatrix}\omega_{il} \\ \omega_{iu} \\ \omega_{jl} \\ \omega_{ju} \\ \omega_{kl} \\ \omega_{ku}\end{pmatrix} = \begin{pmatrix}0\\0\\0\\0\\0\\0\end{pmatrix} \tag{8.35}$$

where $0 < \omega_{il} \leq \omega_{iu} \leq 1; 0 < \omega_{jl} \leq \omega_{ju} \leq 1; 0 < \omega_{kl} \leq \omega_{ku} \leq 1$.

For the reason that $x_{ij} y_{ij} x_{jk} y_{jk} x_{ji} y_{ji} = 1$, we can readily prove that $|\boldsymbol{A}| = 0$ and $\mathrm{R}(\boldsymbol{A}) = 5$. That means that Eqs. (8.35) has nonzero solutions. Therefore, for any given ω_{il} value, $0 < \omega_{il} \leq 1$, we have a solution to Eqs. (8.35).

8.5.3 Priority Model of Inconsistent IFPRs

Consider the equivalent matrices $A = (a_{ij})_{n\times n} = ([\mu_{ij}, p_{ij}])_{n\times n}$ and $B = (b_{ij})_{n\times n} = ([v_{ij}, q_{ij}])_{n\times n}$ of the given IFPR $R = (r_{ij})_{n\times n} = (\mu_{ij}, v_{ij}, \pi_{ij})_{n\times n}$, for which the conditions in Eqs. (8.2), (8.3), and (8.4) hold true. Let $\boldsymbol{\omega} = (\omega_1, \omega_2, \cdots, \omega_n)^T$ be the priority vector of the inconsistent IFNCPR

$A = (a_{ij})_{n\times n} = ([\mu_{ij}, p_{ij}])_{n\times n}$. Obviously, in this case, Eqs. (8.28) and (8.29) do not hold true. So, we introduce the following deviation functions:

$$\begin{cases} g_{ijl} = [\mu_{ij}(\omega_{il} + \omega_{ju}) - \omega_{il}]^2, i, j \in N \\ g_{iju} = [p_{ij}(\omega_{iu} + \omega_{jl}) - \omega_{iu}]^2, i, j \in N \end{cases}$$

It is clear that the smaller values the deviation functions take, the better the consistency of the decision maker's judgment, and that there is no preference relation between the values of the deviation functions g_{ijl} and g_{iju}, for any $i, j \in N$. In order to obtain the priority vector Ω of A, we minimize the deviation functions by constructing the following optimization model.

$$\begin{aligned} min \quad & J = \sum_{i=1}^{n}\sum_{i<j}^{n} [\mu_{ij}(\omega_{il} + \omega_{ju}) - \omega_{il}]^2 + [p_{ij}(\omega_{iu} + \omega_{jl}) - \omega_{iu}]^2 \\ s.t. \quad & 0 \le \omega_{il} \le \omega_{iu} \le 1, \quad i \in N \end{aligned} \tag{8.36}$$

Let $V = (v_1, v_2, \cdots, v_n)^T$ be the priority vector of the inconsistent IFNCPR $B = (b_{ij})_{n\times n} = ([v_{ij}, q_{ij}])_{n\times n}$. We introduce the following deviation functions:

$$\begin{cases} h_{ijl} = [v_{ij}(v_{il} + v_{ju}) - v_{il}]^2, i, j \in N \\ h_{iju} = [q_{ij}(v_{iu} + v_{jl}) - v_{iu}]^2, i, j \in N \end{cases}$$

Similarly, we construct the non-linear programming model as follows:

$$\begin{aligned} min \quad & W = \sum_{i=1}^{n}\sum_{i<j}^{n} [b_{ij}(v_{il} + v_{ju}) - v_{il}]^2 + [q_{ij}(v_{iu} + v_{jl}) - \omega_{iu}]^2 \\ s.t. \quad & 0 \le v_{il} \le v_{iu} \le 1, \quad i \in N \end{aligned} \tag{8.37}$$

In light of Eq.(8.2), Models (8.36) and (8.37) are equivalent. So these models have the same solutions.

Theorem 8.12. Let $\omega = (\omega_1, \omega_2, \cdots, \omega_n)^T$ be the priority vector of the non-multiplicatively consistent IFNCPR $\boldsymbol{A} = (a_{ij})_{n\times n} = ([\mu_{ij}, p_{ij}])_{n\times n}$ and $\boldsymbol{B} = (b_{ij})_{n\times n} = ([v_{ij}, q_{ij}])_{n\times n}$ derived from either Model (8.36) or Model (8.37), and the conditions in Eqs. (8.2), (8.3), and (8.4) hold true, for all $i, j \in N$. Then the priority vector of $\boldsymbol{R} = (r_{ij})_{n\times n} = (\mu_{ij}, v_{ij}, \pi_{ij})_{n\times n}$ is given below:

$$\zeta = ((\omega_{1l}, 1-\omega_{1u}, \omega_{1u} - \omega_{1l}) \ldots (\omega_{il}, 1-\omega_{iu}, \omega_{iu} - \omega_{il}) \ldots (\omega_{nl}, 1-\omega_{nu}, \omega_{nu} - \omega_{nl}))^T$$

Proof. We can show this result by employing the established relationship among $\mu_{ij}, \nu_{ij}, p_{ij}, \pi_{ij}$. QED.

According to Theorems 8.11 and 8.12, the priority vector of an IFPR can be derived by its equivalent matrices.

8.5.4 Priority Algorithm of IFPRs

The priority algorithm of the given inconsistent IFPR $R=(\mu_{ij},\nu_{ij},\pi_{ij})_{n\times n}$ is given below. Due to similarity, the details of the algorithm of a consistent IFPR are omitted.

Step 1: Construct the equivalent matrix $A=(a_{ij})_{n\times n}=([\mu_{ij}, p_{ij}])_{n\times n}$ of the given IFPR $R=(r_{ij})_{n\times n}=(\mu_{ij},\nu_{ij},\pi_{ij})_{n\times n}$.

Step 2: Construct either Model (8.35) or Model (8.36).

Step 3: Solve either Model (8.35) or Model (8.36) as established in Step 2. Let the solution to the model be $\omega_i=(\omega_{il},\omega_{iu}), i\in N$.

Step 4. Obtain the priority vector of the IFPR R as follows:

$$\zeta=((\omega_{1l},1-\omega_{1u},\omega_{1u}-\omega_{1l})\cdots(\omega_{il},1-\omega_{iu},\omega_{iu}-\omega_{il})\ldots\ (\omega_{nl},1-\omega_{nu},\omega_{nu}-\omega_{nl}))^T .$$

Step 5. By using the method of comparing intuitionistic sets as presented in Section 8.2, generate the ranking order of the decision alternatives from the best to the worst.

8.5.5 Numerical Examples

Example 8.3. Consider a set $S=\{s_1,s_2,s_3\}$ of decision alternatives, where the IFPR R_1 is given by the decision maker as follows.

$$\boldsymbol{R}_1=\begin{pmatrix}(0.5,0.5,0) & (0.25,0.5,0.25) & (0.25,0.6,0.15)\\ (0.5,0.25,0.25) & (0.5,0.5,0) & (0.4,0.5,0.1)\\ (0.6,0.25,0.15) & (0.5,0.4,0.1) & (0.5,0.5,0)\end{pmatrix}$$

Step 1: Construct the following equivalent matrix A of R_1.

$$\boldsymbol{A}=\begin{pmatrix}[0.5\ 0.5] & [0.25\ 0.5] & [0.25\ 0.4]\\ [0.5\ 0.75] & [0.5\ 0.5] & [0.4\ 0.5]\\ [0.6\ 0.75] & [0.5\ 0.6] & [0.5\ 0.5]\end{pmatrix}$$

Step 2: According to Eq. (8.18), A is a consistent preference relation of interval fuzzy numbers. The priority vector $\boldsymbol{\omega} = (\omega_1\ \omega_2\ \cdots\ \omega_n)^T = ((\omega_{1l}, \omega_{1u})\ (\omega_{2l}, \omega_{2u})\ (\omega_{3l}, \omega_{3u}))^T$ of A satisfies

$$DS = 0 \tag{8.38}$$

where

$$D = \begin{pmatrix} 3 & 0 & 0 & -1 & 0 & 0 \\ 0 & 1 & -1 & 0 & 0 & 0 \\ 0 & 0 & 3/2 & 0 & 0 & -1 \\ 0 & 0 & 0 & 1 & -1 & 0 \\ 0 & -1 & 0 & 0 & 2/3 & 0 \\ -1 & 0 & 0 & 0 & 0 & 1/3 \end{pmatrix}, \quad S = (\omega_{1l}\ \omega_{1u}\ \omega_{2l}\ \omega_{2u}\ \omega_{3l}\ \omega_{3u})^T .$$

Step 3: Because $|D| = 0$, and $rank(D) = 5$, it can be seen that Eq. (8.38) has solutions. Without loss of generality, let $\omega_{1l} = 0.3$, then we can obtain a solution of Eq. (8.38) as follows:

$$((0.3, 0.4, 0.3), (0.6, 0.1, 0.3), (0.9, 0.1, 0))^T$$

Step 4: The scores of (0.3,0.4,0.3), (0.6,0.1,0.3), and (0.9,0.1,0) are respectively –0.1, 0.5, and 0.8. By using the method of comparing two intuitionistic fuzzy values, the optimal ranking order of the decision alternatives is obtained to be $x_3 \succ x_2 \succ x_1$.

If we let $\omega_{1l} = 0.2$, the other solution to Eq. (8.33) is

$$\boldsymbol{\omega} = (\omega_1, \omega_2, \cdots, \omega_n)^T = ([0.2\ 0.4], [0.4\ 0.6], [0.6\ 0.6])^T$$

So, the priority vector of $\boldsymbol{R}_1$ is $((0.2, 0.8, 0.2), (0.4, 0.4, 0.2), (0.6, 0.4, 0))^T$ .This vector is proportional to the vector obtained above.

Example 8.4. Consider another set $S = \{s_1, s_2, s_3\}$ of decision alternatives. Assume that the IFPR R_2 as given by the decision maker is as follows.

$$\boldsymbol{R}_2 = \begin{pmatrix} (0.5, 0.5, 0) & (0.2, 0.6, 0.2) & (0.6, 0.4, 0) \\ (0.6, 0.2, 0.2) & (0.5, 0.5, 0) & (0.7, 0.1, 0.2) \\ (0.4, 0.6, 0) & (0.1, 0.7, 0.2) & (0.5, 0.5, 0) \end{pmatrix}$$

Step 1: Construct the following equivalent matrix A_2 of R_2.

$$\boldsymbol{A}_2=\begin{pmatrix} [0.5,0.5] & [0.2,0.4] & [0.6,0.6] \\ [0.6,0.8] & [0.5,0.5] & [0.7,0.9] \\ [0.4,0.4] & [0.1,0.3] & [0.5,0.5] \end{pmatrix}$$

Step 2: Because A_2 is an inconsistent preference relation, we construct the following optimization model:

$$\begin{aligned} min\ J = &([0.2(\omega_{1l}+\omega_{2u})-\omega_{1l}]^2+[0.6(\omega_{1l}+\omega_{3u})-\omega_{1l}]^2+[0.7(\omega_{2l}+\omega_{3u})-\omega_{2l}]^2 \\ &+[0.4(\omega_{1u}+\omega_{2l})-\omega_{1u}]^2+[0.6(\omega_{1u}+\omega_{3l})-\omega_{1u}]^2+[0.9(\omega_{2u}+\omega_{3l})-\omega_{2u}]^2) \end{aligned} \tag{8.39}$$

$$s.t.\begin{cases} 0<\omega_{1l}\le\omega_{1u}\le 1, \\ 0<\omega_{2l}\le\omega_{2u}\le 1, \\ 0<\omega_{3l}\le\omega_{3u}\le 1. \end{cases}$$

Step 3: By utilizing the 'Matlab Optimization Toolbox', we obtain the solution to Model (8.39) as follows:

$$\omega_1=[0.2369,0.2369],\omega_2=[0.3565,0.9970],\omega_2=[0.1257,0.1556].$$

Step 4: The priority vector of R_2 is now produced as follows:

$$((0.2369,0.7631,0)(0.3565,0.0030,0.6405)(0.1257,0.8444,0.0299))^T.$$

8.6 Goal Programming Priority Model of IFPRs

In Section 8.5, we discussed a priority method of IFPRs without considering the normalization of intervals. In this section, we will construct a priority model of IFPRs under the normalization restriction of intervals (Sugihara, Ishii and Tanaka, 2004; Gong, Li and Zhou, 2009).

8.6.1 Normalization of Intervals

The method of normalization of intervals was initially proposed by Sugihara, Ishii & Tanaka in 2004 (Sugihara, Ishii and Tanaka, 2004). In particular, the given interval priority vector Ω is said to be normalized if and only if the following conditions are satisfied:

$$\sum_i \omega_{iu}-\max_j(\omega_{ju}-\omega_{jl})\ge 1 \tag{8.40}$$

$$\sum_i \omega_{il}+\max_j(\omega_{ju}-\omega_{jl})\le 1 \tag{8.41}$$

which can be equivalently rewritten as follows:

$$\omega_{il} + \sum_{j=1, j\neq i}^{n} \omega_{iu} \geq 1, i = 1, 2, \ldots, n \tag{8.42}$$

$$\omega_{iu} + \sum_{j=1, j\neq i}^{n} \omega_{il} \leq 1, i = 1, 2, \ldots, n \tag{8.43}$$

8.6.2 Goal Programming Priority Model of Inconsistent IFPRs

In many real-life situations, it is hard for the decision maker to provide any consistent judgment. So, let us consider the equivalent matrices $A = (a_{ij})_{n\times n} = ([\mu_{ij}, p_{ij}])_{n\times n}$ and $B = (b_{ij})_{n\times n} = ([v_{ij}, q_{ij}])_{n\times n}$ of a given inconsistent IFPR $R = (r_{ij})_{n\times n} = (\mu_{ij}, v_{ij}, \pi_{ij})_{n\times n}$, for which the conditions in Eqs. (8.2), (8.3), and (8.4) hold true. Obviously, in this case, A is also inconsistent. If we let $\omega = (\omega_1\ \omega_2\ \cdots\ \omega_n)^T$ be the priority vector of the inconsistent IFNCPR $A = (a_{ij})_{n\times n} = ([\mu_{ij}, p_{ij}])_{n\times n}$, then Eqs. (8.28) and (8.29) will not hold true. In order to get the priority vector of the inconsistent IFPR, we only need to consider the following deviation functions:

$$\varepsilon_{ij} = \mu_{ij}(\omega_{il} + \omega_{ju}) - \omega_{il}, i, j \in N \tag{8.44}$$

$$\gamma_{ij} = p_{ij}(\omega_{iu} + \omega_{jl}) - \omega_{iu}, i, j \in N \tag{8.45}$$

Obviously, the smaller the absolute values the deviation functions take, the better the consistency of the judgment. If $\varepsilon_{ij} = \gamma_{ij} = 0$, we have the conclusion that R is multiplicatively consistent. However, this condition can hardly hold true in real-life situations. In consequence, we can regard $\varepsilon_{ij} = 0$ and $\gamma_{ij} = 0$ as two different goals. It is well known that the goal programming method (Lee, 1973) is especially suitable for dealing with multiple conflicting objectives, and the goal programming allows a simultaneous solution of a system of complex conflicting objectives rather than a single, simple objective as required in linear programming models. Thus we introduce the following goal programming model:

$$min J = \sum_{i=1}^{n}\sum_{j=1}^{n}(|\varepsilon_{ij}| + |\gamma_{ij}|)$$

$$s.t. \begin{cases} \mu_{ij}(\omega_{il} + \omega_{ju}) - \omega_{il} - \varepsilon_{ij} = 0, & i, j \in N; \\ p_{ij}(\omega_{iu} + \omega_{jl}) - \omega_{iu} - \gamma_{ij} = 0, & i, j \in N; \\ \omega_{il} + \sum_{j=1, j\neq i}^{n} \omega_{ju} \geq 1, & i \in N; \\ \omega_{iu} + \sum_{j=1, j\neq i}^{n} \omega_{jl} \leq 1, & i \in N; \\ \omega_{iu} - \omega_{il} \geq 0, & i \in N; \\ \omega_{iu} \geq 0, \omega_{il} \geq 0, & i \in N. \end{cases} \tag{8.46}$$

According to the goal programming theory, ε_{ij} and $|\varepsilon_{ij}|$ can be respectively expressed as $\varepsilon_{ij}=\varepsilon_{ij}^{+}-\varepsilon_{ij}^{-}$ and $|\varepsilon_{ij}|=\varepsilon_{ij}^{+}+\varepsilon_{ij}^{-}$, where $\varepsilon_{ij}^{+}*\varepsilon_{ij}^{-}=0$ for all $i,j\in N$. Similarly, γ_{ij} and $|\gamma_{ij}|$ can be respectively expressed as $\gamma_{ij}=\gamma_{ij}^{+}-\gamma_{ij}^{-}$ and $|\gamma_{ij}|=\gamma_{ij}^{+}+\gamma_{ij}^{-}$, where $\gamma_{ij}^{+}*\gamma_{ij}^{-}=0$ for all $i,j\in N$. The optimization model (8.46) can then be rewritten as the following goal programming model.

$$
minJ=\sum_{i=1}^{n}\sum_{j=1}^{n}(\varepsilon_{ij}^{+}+\varepsilon_{ij}^{-}+\gamma_{ij}^{+}+\gamma_{ij}^{-})
$$

$$
s.t.\begin{cases}
\mu_{ij}(\omega_{il}+\omega_{ju})-\omega_{il}-\varepsilon_{ij}^{+}+\varepsilon_{ij}^{-}=0, & i,j\in N;\\
p_{ij}(\omega_{iu}+\omega_{jl})-\omega_{iu}-\gamma_{ij}^{+}+\gamma_{ij}^{-}=0, & i,j\in N;\\
\omega_{il}+\sum_{j=1,j\neq i}^{n}\omega_{ju}\geq 1, & i\in N;\\
\omega_{iu}+\sum_{j=1,j\neq i}^{n}\omega_{jl}\leq 1, & i\in N;\\
\omega_{iu}-\omega_{il}\geq 0, & i\in N;\\
\omega_{iu}\geq 0,\omega_{il}\geq 0,\varepsilon_{ij}^{+}\geq 0,\varepsilon_{ij}^{-}\geq 0,\gamma_{ij}^{+}\geq 0,\gamma_{ij}^{-}\geq 0 & i\in N.
\end{cases}
\tag{8.47}
$$

According to Section 8.5, the solution to the optimization model in eq. (8.47) can be rewritten to the form of the priority vector ζ of the inconsistent IFPR $R=(r_{ij})_{n\times n}=(\mu_{ij},v_{ij},\pi_{ij})_{n\times n}$. That is, we have

$$
\zeta=((\omega_{1l},1-\omega_{1u},\omega_{1u}-\omega_{1l})\ldots(\omega_{il},1-\omega_{iu},\omega_{iu}-\omega_{il})\ldots(\omega_{nl},1-\omega_{nu},\omega_{nu}-\omega_{nl}))^{T}
$$

8.6.3 Goal Programming Priority of Collectively Inconsistent IFPRs

Suppose that the pairwise matrices provided by the m decision makers are the IFPRs $R^{k}=(\mu_{ij}^{k},v_{ij}^{k},\pi_{ij}^{k})_{n\times n}$, and the individual weights of the decision makers are ϖ_{k}, where ϖ_{k} is a crisp number and $\sum_{k=1}^{m}\varpi_{k}=1$, $k=1,\cdots,m$.

Let $A^{k}=(a_{ij}^{k})_{n\times n}=([\mu_{ij}^{k},p_{ij}^{k}])_{n\times n}$ and $B^{k}=(b_{ij}^{k})_{n\times n}=([v_{ij}^{k},q_{ij}^{k}])_{n\times n}$ be the equivalent matrices of R^{k} such that the conditions in Eqs. (8.2), (8.3), and (8.4) hold true, where $k=1,\cdots,m$. And let $A=(a_{ij})_{n\times n}=[\mu_{ij},p_{ij}]_{n\times n}$ be the weighted arithmetic average combination of $\boldsymbol{A}^{k},k=1,\ldots,m$. Then $a_{ij}=\sum_{k=1}^{m}\varpi_{k}a_{ij}^{k}$. It is clear to see that, for all $i,j\in N$, we have

$$a_{ii} = [\mu_{ii}, p_{ii}] = [\sum_{k=1}^{m} \varpi_k \mu_{ii}^k, \sum_{k=1}^{m} \varpi_k p_{ii}^k] = [0.5, 0.5] \tag{8.48}$$

$$\mu_{ij} + p_{ji} = \sum_{k=1}^{m} \varpi_k \mu_{ij}^k + \sum_{k=1}^{m} \varpi_k p_{ji}^k = \sum_{k=1}^{m} \varpi_k = 1 \tag{8.49}$$

$$\mu_{ji} + p_{ij} = \sum_{k=1}^{m} \varpi_k p_{ij}^k + \sum_{k=1}^{m} \varpi_k \mu_{ji}^k = \sum_{k=1}^{m} \varpi_k = 1 \tag{8.50}$$

which means that $A = (a_{ij})_{n\times n}$ is an IFNCPR.

Let $B = (b_{ij})_{n\times n} = [\nu_{ij}, q_{ij}]_{n\times n}$ be the weighted arithmetic average combination of $\boldsymbol{B}^k, k = 1, \ldots, m$, then $b_{ij} = \sum_{k=1}^{m} \varpi_k b_{ij}^k$. Obviously, the following conditions hold true:

$$b_{ii} = [\nu_{ii}, q_{ii}] = [0.5, 0.5]; \nu_{ij} + q_{ji} = 1; \nu_{ji} + q_{ij} = 1 \tag{8.51}$$

This shows that $B = (b_{ij})_{n\times n}$ is also an IFNCPR. Similarly, by using Eqs. (8.2), we can readily get

$$p_{ij} + \nu_{ij} = 1, q_{ij} + \mu_{ij} = 1 \tag{8.52}$$

If we let $\pi_{ij} = p_{ij} - \mu_{ij}$, then we have

$$\mu_{ij} = \nu_{ji}, \nu_{ij} = \mu_{ji}, \pi_{ij} = \pi_{ji}, \mu_{ij} + \nu_{ij} + \pi_{ij} = 1, i, j \in N \tag{8.53}$$

This shows that $R = (r_{ij})_{n\times n} = (\mu_{ij}, \nu_{ij}, \pi_{ij})_{n\times n}$ is an IFPR.

According to the discussions above, we can regard A and B as equivalent matrices of R; and meanwhile, R can be considered to be the weighted arithmetic average combination of the IFPRs $\boldsymbol{R}^k, k = 1, 2, \ldots, m$ (It should be pointed out that we have not used the disputed operational laws of intuitionistic sets (Atanassov, 1999).

Let $\boldsymbol{R}^k, k = 1, 2, \ldots, m$ be inconsistent. Obviously, in this case, A is inconsistent as well. If we let $= (\omega_1, \omega_2, \cdots, \omega_n)^T$ be the priority vector of the inconsistent IFNCPR $A = (a_{ij})_{n\times n} = ([\mu_{ij}, p_{ij}])_{n\times n}$, then Eqs. (8.28) and (8.29) will not hold true. That is, we have

$$\sum_{k=1}^{m} \varpi_k \mu_{ij}^k = \mu_{ij} \neq \frac{\omega_{il}}{\omega_{il} + \omega_{ju}}, i, j \in N \tag{8.54}$$

$$\sum_{k=1}^{m} \varpi_k p_{ij}^k = p_{ij} \neq \frac{\omega_{iu}}{\omega_{jl} + \omega_{iu}}, i, j \in N \tag{8.55}$$

Similar to the discussion in Subsection 8.6.2, the following deviation functions need to be considered in order to produce the priority vector of the inconsistent IFPR R:

$$\xi_{ij} = \sum_{k=1}^{m} \varpi_k \mu_{ij}^k (\omega_{il} + \omega_{ju}) - \omega_{il}, i, j \in N \tag{8.56}$$

$$\varsigma_{ij} = \sum_{k=1}^{m} \varpi_k p_{ij}^k (\omega_{iu} + \omega_{jl}) - \omega_{iu}, i, j \in N \tag{8.57}$$

It is clear that the smaller the absolute values of the deviation functions are, the better the consistency of decision maker's judgment. The optimization model is therefore derived as follows:

$$\min J = \sum_{i=1}^{n} \sum_{j=1}^{n} (|\xi_{ij}| + |\varsigma_{ij}|)$$

$$s.t. \begin{cases} \sum_{k=1}^{m} \varpi_k \mu_{ij}^k (\omega_{il} + \omega_{ju}) - \omega_{il} - \xi_{ij} = 0, & i, j \in N; \\ \sum_{k=1}^{m} \varpi_k p_{ij}^k (\omega_{iu} + \omega_{jl}) - \omega_{iu} - \varsigma_{ij} = 0, & i, j \in N; \\ \omega_{il} + \sum_{j=1, j \neq i}^{n} \omega_{ju} \geq 1, & i \in N; \\ \omega_{iu} + \sum_{j=1, j \neq i}^{n} \omega_{jl} \leq 1, & i \in N; \\ \omega_{iu} - \omega_{il} \geq 0, & i \in N; \\ \omega_{iu} \geq 0, \omega_{il} \geq 0, & i \in N. \end{cases} \tag{8.58}$$

Similarly, the optimization model in Eq. (8.58) can be rewritten in the following format by utilizing the goal programming theory:

$$minJ = \sum_{i=1}^{n}\sum_{j=1}^{n}(\xi_{ij}^{+} + \xi_{ij}^{-} + \varsigma_{ij}^{+} + \varsigma_{ij}^{-})$$

$$s.t.\begin{cases} \sum_{k=1}^{m}\varpi_k \mu_{ij}^{k}(\omega_{il} + \omega_{ju}) - \omega_{il} - \xi_{ij}^{+} + \xi_{ij}^{-} = 0, & i,j \in N; \\ \sum_{k=1}^{m}\varpi_k p_{ij}^{k}(\omega_{iu} + \omega_{jl}) - \omega_{iu} - \varsigma_{ij}^{+} + \varsigma_{ij}^{-} = 0, & i,j \in N; \\ \omega_{il} + \sum_{j=1, j\neq i}^{n} \omega_{ju} \geq 1, & i \in N; \\ \omega_{iu} + \sum_{j=1, j\neq i}^{n} \omega_{jl} \leq 1, & i \in N; \\ \omega_{iu} - \omega_{il} \geq 0, & i \in N; \\ \omega_{iu} \geq 0, \omega_{il} \geq 0, \varepsilon_{ij}^{+} \geq 0, \varepsilon_{ij}^{-} \geq 0, \gamma_{ij}^{+} \geq 0, \gamma_{ij}^{-} \geq 0, & i \in N. \end{cases} \tag{8.59}$$

The priority vector ζ of the inconsistent IFPR $R = (r_{ij})_{n\times n} = (\mu_{ij}, v_{ij}, \pi_{ij})_{n\times n}$ can be rewritten as:

$$\zeta = ((\omega_{1l}, 1-\omega_{1u}, \omega_{1u} - \omega_{1l})\ldots(\omega_{il}, 1-\omega_{iu}, \omega_{iu} - \omega_{il})\ldots(\omega_{nl}, 1-\omega_{nu}, \omega_{nu} - \omega_{nl}))^{T}$$

where $\omega_{il}, \omega_{iu}, i \in N$, is the solution to model (8.59).

8.6.4 Numerical Examples

Example 8.5. A decision maker (a potential buyer) plans to buy a house. He has three alternatives (houses) $X = \{x_1, x_2, x_3\}$ to choose from. Taking into consideration of various factors, such as price, size of the house, distance to work, environmental characteristics, the decision maker constructs his IFPR R as follows:

$$\boldsymbol{R} = \begin{pmatrix} (0.5,0.5,0) & (0.2,0.6,0.2) & (0.6,0.4,0) \\ (0.6,0.2,0.2) & (0.5,0.5,0) & (0.7,0.1,0.2) \\ (0.4,0.6,0) & (0.1,0.7,0.2) & (0.5,0.5,0) \end{pmatrix}$$

In this IFPR R, the element $(0.2,0.6,0.2)$ denotes the degree of how much the house x_1 is preferred to the house x_2 is 0.2, the degree of how much the house

x_2 is preferred to the house x_1 is 0.6. In other words, the buyer views the possibility of x_1 being superior to x_2 is in the interval $[0.2, 0.4]$, the possibility of x_2 being superior to x_1 is in the interval $[0.6, 0.8]$. The other elements in R can be interpreted in the same way.

Step 1: Construct the following equivalent matrix A of R:

$$A = \begin{pmatrix} [0.5, 0.5] & [0.2, 0.4] & [0.6, 0.6] \\ [0.6, 0.8] & [0.5, 0.5] & [0.7, 0.9] \\ [0.4, 0.4] & [0.1, 0.3] & [0.5, 0.5] \end{pmatrix}.$$

Step 2: Construct the goal programming model as follows:

$$minJ_1 = \varepsilon_{12}^+ + \varepsilon_{12}^- + \varepsilon_{13}^+ + \varepsilon_{13}^- + \varepsilon_{21}^+ + \varepsilon_{21}^- + +\varepsilon_{23}^+ + \varepsilon_{23}^- + \varepsilon_{31}^+ + \varepsilon_{31}^- + \varepsilon_{32}^+ + \varepsilon_{32}^-$$
$$+\gamma_{12}^+ + \gamma_{12}^- + \gamma_{13}^+ + \gamma_{13}^- + \gamma_{23}^+ + \gamma_{23}^- + \gamma_{21}^+ + \gamma_{21}^- + \gamma_{31}^+ + \gamma_{31}^- + \gamma_{32}^+ + \gamma_{32}^-$$

$$s.t. \begin{cases} 0.2(\omega_{1l} + \omega_{2u}) - \omega_{1l} - \varepsilon_{12}^+ + \varepsilon_{12}^- = 0, 0.6(\omega_{1l} + \omega_{3u}) - \omega_{1l} - \varepsilon_{13}^+ + \varepsilon_{13}^- = 0, \\ 0.7(\omega_{2l} + \omega_{3u}) - \omega_{2l} - \varepsilon_{23}^+ + \varepsilon_{23}^- = 0, 0.6(\omega_{2l} + \omega_{1u}) - \omega_{2l} - \varepsilon_{21}^+ + \varepsilon_{21}^- = 0, \\ 0.4(\omega_{3l} + \omega_{1u}) - \omega_{3l} - \varepsilon_{31}^+ + \varepsilon_{31}^- = 0, 0.1(\omega_{3l} + \omega_{2u}) - \omega_{3l} - \varepsilon_{32}^+ + \varepsilon_{32}^- = 0, \\ 0.4(\omega_{1u} + \omega_{2l}) - \omega_{1u} - \gamma_{12}^+ + \gamma_{12}^- = 0, 0.6(\omega_{1u} + \omega_{3l}) - \omega_{1u} - \gamma_{13}^+ + \gamma_{13}^- = 0, \\ 0.9(\omega_{2u} + \omega_{3l}) - \omega_{2u} - \gamma_{23}^+ + \gamma_{23}^- = 0, 0.8(\omega_{2u} + \omega_{1l}) - \omega_{2u} - \gamma_{21}^+ + \gamma_{21}^- = 0, \\ 0.4(\omega_{3u} + \omega_{1l}) - \omega_{3u} - \gamma_{31}^+ + \gamma_{31}^- = 0, 0.3(\omega_{3u} + \omega_{2l}) - \omega_{3u} - \gamma_{32}^+ + \gamma_{32}^- = 0, \\ \omega_{1l} + \omega_{2u} + \omega_{3u} \geq 1, \omega_{2l} + \omega_{1u} + \omega_{3u} \geq 1, \omega_{3l} + \omega_{1u} + \omega_{2u} \geq 1, \\ \omega_{1u} + \omega_{2l} + \omega_{3l} \leq 1, \omega_{2u} + \omega_{1l} + \omega_{3l} \leq 1, \omega_{3u} + \omega_{1l} + \omega_{2l} \leq 1, \\ \omega_{1u} - \omega_{1l} \geq 0, \omega_{2u} - \omega_{2l} \geq 0, \omega_{3u} - \omega_{3l} \geq 0, \\ \omega_{il} \geq 0, \omega_{iu} \geq 0, \varepsilon_{ij}^+ \geq 0, \varepsilon_{ij}^- \geq 0, \gamma_{ij}^+ \geq 0, \varepsilon_{ij}^+ \geq 0, \varepsilon_{ij}^- \geq 0_{ij}^- \geq 0, i, j \in N. \end{cases} \tag{8.60}$$

Step 3: By utilizing the 'Matlab Optimization Toolbox', we produce the solution to (8.54) as follows:

$$\omega_1 = [0.2161, 0.2763], \omega_2 = [0.5136, 0.6659], \omega_3 = [0.1181, 0.2102].$$

Step 4: The priority vector of R is

$$((0.2161, 0.7237, 0.0602)(0.5136, 0.3341, 0.1523)(0.1181, 0.7898, 0.0921))^T.$$

Step 5: From Section 8.2, we obtain that the score of $(0.2161, 0.7237, 0.0602)$ is -0.5076, the score of $(0.5136, 0.3341, 0.1523)$ is 0.1795, and the score of

$(0.1181, 0.7898, 0.0921)$ is -0.6717. Thus, the optimal ranking order of the decision alternatives is obtained as $x_2 \succ x_1 \succ x_3$.

Example 8.6. A decision maker (a potential buyer) invites three experts to help him buy a house, and the weights for the individual decision makers are respectively 0.3, 0.4, and 0.3. There are three alternatives (houses) $X = \{x_1, x_2, x_3\}$ for the buyer to choose from. The IFPRs R_i, as presented by the ith, $i = 1, 2, 3$, expert, are detailed as follows:

$$\boldsymbol{R}_1 = \begin{pmatrix} (0.5,0.5,0) & (0.1,0.6,0.3) & (0.6,0.3,0.1) \\ (0.6,0.1,0.3) & (0.5,0.5,0) & (0.8,0.2,0) \\ (0.3,0.6,0.1) & (0.2,0.8,0) & (0.5,0.5,0) \end{pmatrix}$$

$$\boldsymbol{R}_2 = \begin{pmatrix} (0.5,0.5,0) & (0.3,0.7,0) & (0.7,0.2,0.1) \\ (0.7,0.3,0) & (0.5,0.5,0) & (0.6,0.2,0.2) \\ (0.2,0.7,0.1) & (0.2,0.6,0.2) & (0.5,0.5,0) \end{pmatrix}$$

$$\boldsymbol{R}_3 = \begin{pmatrix} (0.5,0.5,0) & (0.3,0.6,0.1) & (0.9,0.1,0) \\ (0.6,0.3,0.1) & (0.5,0.5,0) & (0.7,0.2,0.1) \\ (0.1,0.9,0) & (0.2,0.7,0.1) & (0.5,0.5,0) \end{pmatrix}$$

Step 1: Construct the following equivalent matrix A_i of R_i, $i = 1, 2, 3$.

$$\boldsymbol{A}_1 = \begin{pmatrix} [0.5,0.5] & [0.1,0.4] & [0.6,0.7] \\ [0.6,0.9] & [0.5,0.5] & [0.8,0.8] \\ [0.3,0.4] & [0.2,0.2] & [0.5,0.5] \end{pmatrix}$$

$$\boldsymbol{A}_2 = \begin{pmatrix} [0.5,0.5] & [0.3,0.3] & [0.7,0.8] \\ [0.7,0.7] & [0.5,0.5] & [0.6,0.8] \\ [0.2,0.3] & [0.2,0.4] & [0.5,0.5] \end{pmatrix}$$

$$\boldsymbol{A}_3 = \begin{pmatrix} [0.5,0.5] & [0.3,0.4] & [0.9,0.9] \\ [0.6,0.7] & [0.5,0.5] & [0.7,0.8] \\ [0.1,0.1] & [0.2,0.3] & [0.5,0.5] \end{pmatrix}$$

Step 2: The collective IFPR R of the individual $R_i, i = 1, 2, 3$, and its equivalent matrix A are constructed as follows:

$$\boldsymbol{R}=\begin{pmatrix} (0.5,0.5,0) & (0.24,0.64,0.12) & (0.73,0.2,0.07) \\ (0.64,0.24,0.12) & (0.5,0.5,0) & (0.69,0.2,0.11) \\ (0.2,0.73,0.07) & (0.2,0.69,0.11) & (0.5,0.5,0) \end{pmatrix}$$

$$\boldsymbol{A}=\begin{pmatrix} [0.5,0.5] & [0.24,0.36] & [0.73,0.8] \\ [0.64,0.76] & [0.5,0.5] & [0.69,0.8] \\ [0.20,0.27] & [0.2,0.31] & [0.5,0.5] \end{pmatrix}$$

Step 3: Construct the necessary optimal model as follows:

$$min J_2=\xi_{12}^{+}+\xi_{12}^{-}+\xi_{13}^{+}+\xi_{13}^{-}+\xi_{21}^{+}+\xi_{21}^{-}++\xi_{23}^{+}+\xi_{23}^{-}+\xi_{31}^{+}+\xi_{31}^{-}+\xi_{32}^{+}+\xi_{32}^{-}$$
$$+\varsigma_{12}^{+}+\varsigma_{12}^{-}+\varsigma_{13}^{+}+\varsigma_{13}^{-}+\varsigma_{23}^{+}+\varsigma_{23}^{-}+\varsigma_{21}^{+}+\varsigma_{21}^{-}+\varsigma_{31}^{+}+\varsigma_{31}^{-}+\varsigma_{32}^{+}+\varsigma_{32}^{-}$$

$$s.t.\begin{cases} 0.24(\omega_{1l}+\omega_{2u})-\omega_{1l}-\xi_{12}^{+}+\xi_{12}^{-}=0, 0.73(\omega_{1l}+\omega_{3u})-\omega_{1l}-\xi_{13}^{+}+\xi_{13}^{-}=0, \\ 0.69(\omega_{2l}+\omega_{3u})-\omega_{2l}-\xi_{23}^{+}+\xi_{23}^{-}=0, 0.64(\omega_{2l}+\omega_{1u})-\omega_{2l}-\xi_{21}^{+}+\xi_{21}^{-}=0, \\ 0.2(\omega_{3l}+\omega_{1u})-\omega_{3l}-\xi_{31}^{+}+\xi_{31}^{-}=0, 0.2(\omega_{3l}+\omega_{2u})-\omega_{3l}-\xi_{32}^{+}+\xi_{32}^{-}=0, \\ 0.36(\omega_{1u}+\omega_{2l})-\omega_{1u}-\varsigma_{12}^{+}+\varsigma_{12}^{-}=0, 0.8(\omega_{1u}+\omega_{3l})-\omega_{1u}-\varsigma_{13}^{+}+\varsigma_{13}^{-}=0, \\ 0.8(\omega_{2u}+\omega_{3l})-\omega_{2u}-\varsigma_{23}^{+}+\varsigma_{23}^{-}=0, 0.76(\omega_{2u}+\omega_{1l})-\omega_{2u}-\varsigma_{21}^{+}+\varsigma_{21}^{-}=0, \\ 0.27(\omega_{3u}+\omega_{1l})-\omega_{3u}-\varsigma_{31}^{+}+\varsigma_{31}^{-}=0, 0.31(\omega_{3u}+\omega_{2l})-\omega_{3u}-\varsigma_{32}^{+}+\varsigma_{32}^{-}=0, \\ \omega_{1l}+\omega_{2u}+\omega_{3u}\ge 1, \omega_{2l}+\omega_{1u}+\omega_{3u}\ge 1, \omega_{3l}+\omega_{1u}+\omega_{2u}\ge 1, \\ \omega_{1u}+\omega_{2l}+\omega_{3l}\le 1, \omega_{2u}+\omega_{1l}+\omega_{3l}\le 1, \omega_{3u}+\omega_{1l}+\omega_{2l}\le 1, \\ \omega_{1u}-\omega_{1l}\ge 0, \omega_{2u}-\omega_{2l}\ge 0, \omega_{3u}-\omega_{3l}\ge 0, \\ \omega_{il}\ge 0, \omega_{iu}\ge 0, \varepsilon_{ij}^{+}\ge 0, \varepsilon_{ij}^{-}\ge 0, \gamma_{ij}^{+}\ge 0, \varepsilon_{ij}^{+}\ge 0, \varepsilon_{ij}^{-}\ge 0_{ij}^{-}\ge 0, i,j\in N. \end{cases} \quad (8.61)$$

Step 4: By utilizing the 'Matlab Optimization Toolbox', we obtain the solution to Eq. (8.61) as follows:

$$\omega_1=[0.2599,0.3481], \omega_2=[0.4946,0.6368], \omega_3=[0.1033,0.1574]$$

Step 5: The priority vector of R is now produced as follows:

$$((0.2599,0.6519,0.0882)(0.4946,0.3632,0.1422)(0.1033,0.8426,0.0541))^T.$$

Step 6: By using the method of comparing intuitionistic fuzzy values, the optimal ranking order of the alternatives is obtained as $x_2 \succ x_1 \succ x_3$.

8.7 Optimal Priority Models of Additively Consistent IFPRs

Basing on the results of additively consistent IFPRs, as presented in Section 8.3, we now discuss the priority methods of the IFPRs (Gong, Li, Forrest and Zhao, 2011).

8.7.1 Additively Consistent IFPRs

According to Section 3.2, an IFNCPR $R' = (r'_{ij})_{n\times n}$ is additively consistent if there exists a vector $\boldsymbol{V} = (v_1 \ldots v_n)^T = ([v_{1l}, v_{1u}] \ldots [v_{nl}, v_{nu}])^T$ such that

$$r'_{ij} = 0.5 + 0.2log3^{v_i/v_j} = [0.5 + 0.2log3^{v_{il}/v_{ju}}, 0.5 + 0.2log3^{v_{iu}/v_{jl}}] \quad \forall i, j \in N,$$

where V and $v_i = [v_{il}, v_{iu}], i \in N$, are called the priority vector of the additively consistent IFNPR R' and the weight of $s_i, i \in N$, respectively. In fact, $v_i, i \in N$, can also be interpreted as the range of the membership degree of the importance of the decision alternative $s_i, i \in N$.

Consider the equivalent matrices $A = (a_{ij})_{n\times n} = ([\mu_{ij}, p_{ij}])_{n\times n}$ and $B = (b_{ij})_{n\times n} = ([v_{ij}, q_{ij}])_{n\times n}$ of the IFPR $R = (r_{ij})_{n\times n} = (\mu_{ij}, v_{ij}, \pi_{ij})_{n\times n}$, for which the conditions in Eqs. (8.2), (8.3), and (8.4) hold true. Let $\boldsymbol{V} = (v_1 \ldots v_n)^T = ([v_{1l}, v_{1u}] \ldots [v_{nl}, v_{nu}])^T$ be the priority vector of the additively consistent IFNPR A, then we have

$$a_{ij} = [\mu_{ij}, p_{ij}] = 0.5 + 0.2log3^{v_i/v_j} = [0.5 + 0.2log3^{v_{il}/v_{ju}}, 0.5 + 0.2log3^{v_{iu}/v_{jl}}], i, j \in N \quad (8.62)$$

That is, we have

$$\mu_{ij} = 0.5 + 0.2log3^{v_{il}/v_{ju}}, i, j \in N \quad (8.63)$$

$$p_{ij} = 0.5 + 0.2log3^{v_{iu}/v_{jl}}, i, j \in N \quad (8.64)$$

According to Eqs. (8.2), we have

$$v_{ij} = 1 - p_{ij} = 0.5 + 0.2log3^{v_{jl}/v_{iu}}, i, j \in N \quad (8.65)$$

$$q_{ij} = 1 - \mu_{ij} = 0.5 + 0.2log3^{v_{ju}/v_{il}}, i, j \in N \quad (8.66)$$

Thus $b_{ij} = [v_{ij}, q_{ij}] = 0.5 + 0.2log3^{v_j/v_i} = [0.5 + 0.2log3^{v_{jl}/v_{iu}}, 0.5 + 0.2log3^{v_{ju}/v_{il}}], i, j \in N$. This result shows that $\boldsymbol{V} = (v_1 \ldots v_n)^T$ is also the priority vector of $\boldsymbol{B} = (b_{ij})_{n\times n} = ([v_{ij}, q_{ij}])_{n\times n}$. Therefore, we have the following result.

Theorem 8.13. If the additively consistent IFNCPR $A = (a_{ij})_{n\times n} = ([\mu_{ij}, p_{ij}])_{n\times n}$ and $B = (b_{ij})_{n\times n} = ([v_{ij}, q_{ij}])_{n\times n}$ satisfy Eqs. (8.2), (8.3), and (8.4), then the matrices A and B have the same priority vector.

According to Eqs. (8.63) and (8.64), or Eqs. (8.65) and (8.66), we can produce

$$\pi_{ij} = p_{ij} - \mu_{ij} = 0.2log3^{v_{iu}v_{ju}/v_{il}v_{jl}}, i, j \in N \tag{8.67}$$

Let $V = (v_1 \ldots v_n)^T = ([v_{1l}, v_{1u}] \ldots [v_{nl}, v_{nu}])^T$ be the priority vector of the additively consistent IFNCPR A, where $v_i = [v_{il}, v_{iu}]$ is the range of the membership degree of the importance of the decision alternative s_i. Then we refer $1 - v_i = [1 - v_{iu}, 1 - v_{il}]$ to as the range of the non-membership degree of the importance of the decision alternative s_i, and $v_{iu} - v_{il}$ the hesitation degree of the importance of the alternative s_i, $i \in N$. Thus, we can construct the priority vector of the IFPR R in the following form of the intuitionistic fuzzy numbers:

$$\vartheta = ((v_{1l}, 1 - v_{1u}, v_{1u} - v_{1l}) \ldots (v_{il}, 1 - v_{iu}, v_{iu} - v_{il}) \ldots (v_{nl}, 1 - v_{nu}, v_{nu} - v_{nl}))^T,$$

where v_{il} can be interpreted as the membership degree of the importance (weight) of s_i, $1 - v_{iu}$ the non-membership degree of the importance (weight) of s_i, and $v_{iu} - v_{il}$ the hesitation degree of the importance (weight) of s_i, $i \in N$.

Therefore, Eqs. (8.63), (8.65), and (8.67) jointly mean that the membership degree μ_{ij}, the non-membership degree v_{ij}, and the intuitionistic fuzzy index π_{ij} of R can be represented by its priority vector ϑ. Thus we naturally develop the concept of additively consistent IFPRs as follows.

Definition 8.20. An IFPR $R = (r_{ij})_{n \times n}$ is additively consistent, if there exists a vector $\vartheta = (\vartheta_1 \ldots \vartheta_n)^T$ such that Eqs. (8.63), (8.65), and (8.67) hold true, where $\vartheta_i = (v_{il}, 1 - v_{iu}, v_{iu} - v_{il}), i \in N$, and ϑ is called the intuitionistic fuzzy priority vector of the additively consistent IFPR R. Additionally, the conditions in Eqs. (8.63), (8.65), and (8.67) are referred to as the additive consistency conditions of *R*.

For simplicity, we denote the interval vector $V = (v_1 \ldots v_n)^T = ([v_{1l}, v_{1u}] \ldots [v_{nl}, v_{nu}])^T$ as $V' = (v_{1l}\ v_{1u} \ldots v_{nl}\ v_{nu})^T$, which is known as the 'characteristics vector' of R. Consider the fact that the intuitionistic fuzzy vector

$$\vartheta = ((v_{1l}, 1 - v_{1u}, v_{1u} - v_{1l}) \ldots (v_{il}, 1 - v_{iu}, v_{iu} - v_{il}) \ldots (v_{nl}, 1 - v_{nu}, v_{nu} - v_{nl}))^T$$

derives from the interval vector V, we can also construct ϑ by 'characteristics vector' V'.

Let us reconsider Eqs. (8.63), (8.65), and (8.67). As a matter of fact, Eq. (8.67) actually originates from Eqs. (8.63) and (8.64), and Eq. (8.65) originates from Eq. (8.64). Therefore, in order to get the priority order of a given IFPR, we only need to take Eqs. (8.63) and (8.65) into account. That is, we have

$$\mu_{ij} = 0.5 + 0.2log3^{v_{il}/v_{ju}}; 1 - v_{ij} = p_{ij} = 0.5 + 0.2log3^{v_{iu}/v_{jl}}, i, j \in N. \tag{8.68}$$

If there is no priority vector $\vartheta = (\vartheta_1 \ldots \vartheta_n)^T$ such that Eqs. (8.63) and (8.64) hold true, then we say that R is not additively consistent or additively inconsistent.

Theorem 8.14. Let $\boldsymbol{\theta} = (\theta_1 \ldots \theta_n)^T$ be the priority vector of the given IFNCPRs $\boldsymbol{A} = (a_{ij})_{n\times n} = ([\mu_{ij}, p_{ij}])_{n\times n}$ and $\boldsymbol{B} = (b_{ij})_{n\times n} = ([v_{ij}, q_{ij}])_{n\times n}$. And assume that the conditions in Eqs. (8.2), (8.3), and (8.4) hold true for all $i, j \in N$. Then the priority vector of $\boldsymbol{R} = (r_{ij})_{n\times n} = (\mu_{ij}, v_{ij}, \pi_{ij})_{n\times n}$ is given as follows:

$$\boldsymbol{\theta} = ((v_{1l}, 1 - v_{1u}, v_{1u} - v_{1l}) \ldots (v_{il}, 1 - v_{iu}, v_{iu} - v_{il}) \ldots (v_{nl}, 1 - v_{nu}, v_{nu} - v_{nl}))^T.$$

Proof. This result can be shown by using the relationship among $\mu_{ij}, v_{ij}, p_{ij}, \pi_{ij}$. All the details are omitted. QED.

8.7.2 *Priority of Collective Additively Consistent IFPRs*

Suppose that the IFPRs provided by m decision makers are $R^k = (\mu_{ij}^k, v_{ij}^k, \pi_{ij}^k)_{n\times n}, k = 1, 2, \ldots, m$, and the weight for decision maker k is ϖ_k satisfying $\sum_{k=1}^{m} \varpi_k = 1$, $k = 1, \cdots, m$. Let $\mu_{ij} = \sum_{k=1}^{m} \varpi_k \mu_{ij}^k$, $v_{ij} = \sum_{k=1}^{m} \varpi_k v_{ij}^k$, and $\pi_{ij} = \sum_{k=1}^{m} \varpi_k \pi_{ij}^k, i, j \in N$. Then, for $\mu_{ij}^k = v_{ji}^k, v_{ij}^k = \mu_{ji}^k, \pi_{ij}^k = \pi_{ji}^k; \mu_{ij}^k + v_{ij}^k + \pi_{ij}^k = 1$, we have

$$\sum_{k=1}^{m} \varpi_k \mu_{ij}^k = \sum_{k=1}^{m} \varpi_k v_{ji}^k \tag{8.69}$$

$$\sum_{k=1}^{m} \varpi_k v_{ij}^k = \sum_{k=1}^{m} \varpi_k \mu_{ji}^k \tag{8.70}$$

$$\sum_{k=1}^{m} \varpi_k \pi_{ij}^k = \sum_{k=1}^{m} \varpi_k \pi_{ji}^k \tag{8.71}$$

$$\sum_{k=1}^{m} \varpi_k \mu_{ij}^k + \sum_{k=1}^{m} \varpi_k v_{ij}^k + \sum_{k=1}^{m} \varpi_k \pi_{ij}^k = \sum_{k=1}^{m} \varpi_k = 1 \tag{8.72}$$

If we let $\mu_{ij} = \sum_{k=1}^{m} \varpi_k \mu_{ij}^k$, $v_{ij} = \sum_{k=1}^{m} \varpi_k v_{ij}^k$, $\pi_{ij} = \sum_{k=1}^{m} \varpi_k \pi_{ij}^k \forall i, j \in N$, then Eqs. (8.69) - (8.72) are equivalent to the following:

$$\mu_{ij} = v_{ji}, v_{ij} = \mu_{ji}, \pi_{ij} = \pi_{ji}; \mu_{ij} + v_{ij} + \pi_{ij} = 1 \quad (8.73)$$

Similarly, we can also readily produce

$$(\mu_{ii}, v_{ii}, \pi_{ii}) = (0.5, 0.5, 0) \quad (8.74)$$

Therefore, $R = (r_{ij})_{n\times n} = (\mu_{ij}, v_{ij}, \pi_{ij})_{n\times n}$ can be regarded as the weighted arithmetic average combination of the IFPRs $R = (r_{ij}^k)_{n\times n} = (\mu_{ij}^k, v_{ij}^k, \pi_{ij}^k)_{n\times n}$, $k = 1, 2, \ldots, m$; and it is still an IFPR in terms of Eqs. (8.73) and (8.74).

Let $V' = (v_{1l}\ v_{1u} \ldots v_{nl}\ v_{nu})^T$ and $V'_k = (v_{1l}^k\ v_{1u}^k \ldots v_{nl}^k\ v_{nu}^k)^T$ be respectively 'the characteristic vectors' of the additively consistent IFPRs R and $R_k, k = 1, 2, \ldots, m$. In the following, we will discuss the relationship between V' and $V'_k, k = 1, 2, \ldots, m, i \in N$. To this end, according to Eqs. (8.63) and (8.64), we have

$$\mu_{ij} = 0.5 + 0.2 log 3^{v_{il}/v_{ju}}; 1 - v_{ij} = p_{ij} = 0.5 + 0.2 log 3^{v_{iu}/v_{jl}}, i, j \in N \quad (8.75)$$

$$\mu_{ij}^k = 0.5 + 0.2 log 3^{v_{il}^k/v_{ju}^k}; 1 - v_{ij}^k = p_{ij}^k = 0.5 + 0.2 log 3^{v_{iu}^k/v_{jl}^k}, i, j \in N \quad (8.76)$$

For $\mu_{ij} = \sum_{k=1}^{m} \varpi_k \mu_{ij}^k$, we then have

$$0.5 + 0.2 log 3^{v_{il}/v_{ju}} = \sum_{k=1}^{m} \varpi_k (0.5 + 0.2 log 3^{v_{il}^k/v_{ju}^k}) = 0.5 + 0.2 log 3^{\prod_{k=1}^{m} (v_{il}^k/v_{ju}^k)^{\varpi_k}} \quad (8.77)$$

Thus we have

$$v_{il} / v_{ju} = \prod_{k=1}^{m} (v_{il}^k / v_{ju}^k)^{\varpi_k} \quad (8.78)$$

For $v_{ij} = \sum_{k=1}^{m} \varpi_k v_{ij}^k$, we have

$$0.5 + 0.2 log 3^{v_{iu}/v_{jl}} = \sum_{k=1}^{m} \varpi_k (0.5 + 0.2 log 3^{v_{iu}^k/v_{jl}^k}) = 0.5 + 0.2 log 3^{\prod_{k=1}^{m} (v_{iu}^k/v_{jl}^k)^{\varpi_k}} \quad (8.79)$$

Thus we have

$$v_{iu} / v_{jl} = \prod_{k=1}^{m} (v_{iu}^k / v_{jl}^k)^{\varpi_k} \quad (8.80)$$

In consequence, the 'the characteristic vectors' V' and V'_k, $k = 1, 2, \ldots, m$, satisfy the relations as given in Eqs. (8.78) and (8.80).

8.7.3 The Priority of an Individual Inconsistent IFPR

In many real-life situations, it is hard for the decision maker to provide a consistent judgment. So, to study such situations, we let

$$\theta = ((v_{1l}, 1-v_{1u}, v_{1u}-v_{1l})\ldots(v_{il}, 1-v_{iu}, v_{iu}-v_{il})\ldots(v_{nl}, 1-v_{nu}, v_{nu}-v_{nl}))^T$$

be the priority vector of the given IFPR $R=(r_{ij})_{n\times n}=(\mu_{ij},v_{ij},\pi_{ij})_{n\times n}$. By using Eq. (8.68), we have

$$v_{il}-\mu'_{ij}v_{ju}=0; v_{iu}-v'_{ij}v_{jl}=0, i, j\in N \tag{8.81}$$

where $\mu'_{ij}=3^{5(\mu_{ij}-0.5)}, v_{ij}=3^{5(0.5-v_{ij})}$. Obviously, if R is inconsistent, then Eqs. (8.81) can't hold true. For such situations, let us introduce the following deviation functions:

$$\varepsilon_{ij}=v_{il}-\mu'_{ij}v_{ju} \tag{8.82}$$

$$\gamma_{ij}=v_{iu}-v'_{ij}v_{jl} \tag{8.83}$$

It is ready to see that the smaller the squared values or the absolute values of the deviation functions are, the better the consistency of judgment is. In the following, we will introduce the least squares optimization model and the goal programming optimization model to solve for the priority vector of R.

The least squares optimization model:

$$min\ J_1=\sum_{i=1}^{n}\sum_{j=1}^{n}(v_{il}-\mu'_{ij}v_{ju})^2+(v_{iu}-v'_{ij}v_{jl})^2$$

$$s.t.\begin{cases} v_{il}+\sum_{j=1, j\neq i}^{n} v_{ju}\geq 1, & i\in N;\\ v_{iu}+\sum_{j=1, j\neq i}^{n} v_{jl}\leq 1, & i\in N;\\ v_{iu}-v_{il}\geq 0, & i\in N;\\ v_{iu}\geq 0, v_{il}\geq 0, & i\in N. \end{cases} \tag{8.84}$$

The goal programming optimization model:

$$\min J_2 = \sum_{i=1}^{n}\sum_{j=1}^{n}(|\varepsilon_{ij}|+|\gamma_{ij}|)$$

$$s.t.\begin{cases} v_{il} - \mu'_{ij}v_{ju} - \varepsilon_{ij} = 0, & i,j \in N; \\ v_{iu} - v'_{ij}v_{jl} - \gamma_{ij} = 0, & i,j \in N; \\ v_{il} + \sum_{j=1, j\neq i}^{n} v_{ju} \geq 1, & i \in N; \\ v_{iu} + \sum_{j=1, j\neq i}^{n} v_{jl} \leq 1, & i \in N; \\ v_{iu} - v_{il} \geq 0, & i \in N; \\ v_{iu} \geq 0, v_{il} \geq 0, & i \in N. \end{cases} \tag{8.85}$$

According to the goal programming theory, the optimization model in Eq. (8.85) can be rewritten as follows:

$$\min J_2 = \sum_{i=1}^{n}\sum_{j=1}^{n}(\varepsilon_{ij}^{+}+\varepsilon_{ij}^{-}+\gamma_{ij}^{+}+\gamma_{ij}^{-})$$

$$s.t.\begin{cases} v_{il} - \mu'_{ij}v_{ju} - \varepsilon_{ij}^{+} + \varepsilon_{ij}^{-} = 0, \\ v_{iu} - v'_{ij}v_{jl} - \gamma_{ij}^{+} + \gamma_{ij}^{-} = 0, \\ v_{il} + \sum_{j=1, j\neq i}^{n} v_{ju} \geq 1, \\ v_{iu} + \sum_{j=1, j\neq i}^{n} v_{jl} \leq 1, \\ v_{iu} - v_{il} \geq 0, \\ v_{iu} \geq 0, v_{il} \geq 0, \varepsilon_{ij}^{+} \geq 0, \varepsilon_{ij}^{-} \geq 0, \gamma_{ij}^{+} \geq 0, \gamma_{ij}^{-} \geq 0, \\ i,j \in N. \end{cases} \tag{8.86}$$

It should be pointed out that the constraints $v_{il} + \sum_{j=1, j\neq i}^{n} v_{ju} \geq 1$ and $v_{iu} + \sum_{j=1, j\neq i}^{n} v_{jl} \leq 1$ represent the normalization of the interval vector V, for details, see (Sugihara et al., 2006; Wang, and Elhag, 2006).

8.7.4 Priority of Collective Inconsistent IFPRs

Suppose that the IFPRs provided by the m decision makers are $R^k = (\mu_{ij}^k, v_{ij}^k, \pi_{ij}^k)_{n\times n}, k = 1,2,\ldots,m$, and the weight for the kth decision maker is ϖ_k satisfying $\sum_{k=1}^{m}\varpi_k = 1$, $k = 1,\cdots,m$. Let $R = (r_{ij})_{n\times n} = (\mu_{ij}, v_{ij}, \pi_{ij})$ be the weighted arithmetic average combination of $R^k, k = 1,2,\ldots,m$. Then we have $\mu_{ij} = \sum_{k=1}^{m}\varpi_k\mu_{ij}^k$, $v_{ij} = \sum_{k=1}^{m}\varpi_k v_{ij}^k$, and $\pi_{ij} = \sum_{k=1}^{m}\varpi_k\pi_{ij}^k, i, j\in N$.

Obviously, R is an IFPR. Suppose that $R^k, k = 1,2,\ldots,m$, are respectively inconsistent. Then it is clear that R is also inconsistent. If we let

$$\theta = ((v_{1l}, 1-v_{1u}, v_{1u}-v_{1l})\ldots(v_{il}, 1-v_{iu}, v_{iu}-v_{il})\ldots(v_{nl}, 1-v_{nu}, v_{nu}-v_{nl}))^T$$

be the priority vector of the additively consistent IFNPR R, then by using Eqs. (8.68), we have

$$\sum_{k=1}^{m}\varpi_k\mu_{ij}^k = \mu_{ij} = 0.5 + 0.2log3^{v_{il}/v_{ju}}, i, j\in N \tag{8.87}$$

$$\sum_{k=1}^{m}\varpi_k(1-v_{ij}^k) = 1-v_{ij} = 0.5 + 0.2log3^{v_{iu}/v_{jl}}, i, j\in N \tag{8.88}$$

And, so Eqs.(8.87) and (8.88) are equivalent to the following:

$$v_{il} - \mu''_{ij}v_{ju} = 0, i, j\in N \tag{8.89}$$

$$v_{iu} - v''_{ij}v_{jl} = 0, i, j\in N \tag{8.90}$$

where $\mu''_{ij} = 3^{5(\sum_{k=1}^{m}\varpi_k\mu_{ij}^k - 0.5)}$ and $v''_{ij} = 3^{5(0.5-\sum_{k=1}^{m}\varpi_k v_{ij}^k)}$. Obviously, if R is inconsistent, then Eqs. (8.89) and (8.90) can't hold true. The deviation functions are now constructed as follows:

$$\xi_{ij} = v_{il} - \mu''_{ij}v_{ju}, i, j\in N \tag{8.91}$$

$$\varsigma_{ij} = v_{iu} - v''_{ij}v_{jl}, i, j\in N \tag{8.92}$$

It can be seen that the smaller the squares values or the absolute values of the deviation functions are, the better the consistency of the judgment is. The least squares optimization priority model and the goal programming optimization priority model of the collective inconsistent IFPRs are constructed as follows:

The least squares optimization model:

$$min\ J=\sum_{i=1}^{n}\sum_{j=1}^{n}(v_{il}-\mu''_{ij}v_{ju})^2+(v_{iu}-v''_{ij}v_{jl})^2$$

$$s.t.\begin{cases} v_{il}+\sum\limits_{j=1,j\neq i}^{n} v_{ju}\geq 1, & i\in N;\\ v_{iu}+\sum\limits_{j=1,j\neq i}^{n} v_{jl}\leq 1, & i\in N;\\ v_{iu}-v_{il}\geq 0, & i\in N;\\ v_{iu}\geq 0, v_{il}\geq 0, & i\in N. \end{cases} \tag{8.93}$$

The goal programming optimization model:

$$minJ=\sum_{i=1}^{n}\sum_{j=1}^{n}(\xi_{ij}^{+}+\xi_{ij}^{-}+\varsigma_{ij}^{+}+\varsigma_{ij}^{-})$$

$$s.t.\begin{cases} v_{il}-\mu''_{ij}v_{ju}-\xi_{ij}^{+}+\xi_{ij}^{-}=0,\\ v_{iu}-v''_{ij}v_{jl}-\varsigma_{ij}^{+}+\varsigma_{ij}^{-}=0,\\ v_{il}+\sum\limits_{j=1,j\neq i}^{n} v_{ju}\geq 1,\\ v_{iu}+\sum\limits_{j=1,j\neq i}^{n} v_{jl}\leq 1,\\ v_{iu}-v_{il}\geq 0,\\ v_{iu}\geq 0, v_{il}\geq 0, \xi_{ij}^{+}\geq 0, \xi_{ij}^{-}\geq 0, \varsigma_{ij}^{+}\geq 0, \varsigma_{ij}^{-}\geq 0,\\ i,j\in N. \end{cases} \tag{8.94}$$

where $\mu''_{ij}=3^{5(\sum_{k=1}^{m}\varpi_k\mu_{ij}^{k}-0.5)}$ and $v''_{ij}=3^{5(0.5-\sum_{k=1}^{m}\varpi_k v_{ij}^{k})}$, $i,j\in N$.

8.7.5 Numerical Examples

Example 8.7. Consider the set $X=\{x_1,x_2,x_3\}$ of decision alternatives. Assume that the IFPR R as given by the decision maker is as follows:

$$R=\begin{pmatrix} (0.5,0.5,0) & (0.2,0.6,0.2) & (0.6,0.4,0)\\ (0.6,0.2,0.2) & (0.5,0.5,0) & (0.7,0.1,0.2)\\ (0.4,0.6,0) & (0.1,0.7,0.2) & (0.5,0.5,0) \end{pmatrix}.$$

The least squares optimization model:

Step 1: Construct the least squares optimization model as follows:

$$minJ = \sum_{i=1}^{n}\sum_{j=1}^{n}(v_{1l}-3^{-1.5}v_{2u})^2+(v_{1l}-3^{0.5}v_{3u})^2+(v_{2l}-3^{1.0}v_{3u})^2+(v_{2l}-3^{0.5}v_{1u})^2$$
$$+(v_{3l}-3^{-0.5}v_{1u})^2+(v_{3l}-3^{-2.0}v_{2u})^2+(v_{1u}-3^{-0.5}v_{2l})^2+(v_{1u}-3^{0.5}v_{3l})^2$$
$$+(v_{2u}-3^{2.0}v_{3l})^2+(v_{2u}-3^{1.5}v_{1l})^2+(v_{3u}-3^{-0.5}v_{1l})^2+(v_{3u}-3^{-1.0}v_{2l})^2 \quad (8.95)$$
$$s.t.\begin{cases} v_{1l}+v_{2u}+v_{3u}\geq 1, v_{2l}+v_{1u}+v_{3u}\geq 1, v_{3l}+v_{1u}+v_{2u}\geq 1, \\ v_{1u}+v_{2l}+v_{3l}\leq 1, v_{2u}+v_{1l}+v_{3l}\leq 1, v_{3u}+v_{1l}+v_{2l}\leq 1, \\ v_{1u}-v_{1l}\geq 0, v_{2u}-v_{2l}\geq 0, v_{3u}-v_{3l}\geq 0, \\ v_{il}\geq 0, v_{iu}\geq 0, \varepsilon_{ij}^{+}\geq 0, \varepsilon_{ij}^{-}\geq 0, \gamma_{ij}^{+}\geq 0, \gamma_{ij}^{-}\geq 0, i,j\in N. \end{cases}$$

Step 2: By utilizing 'Matlab Optimization Toolbox', we get the following solution to Eq. (8.95):

$$v_{1l}=0.1508, v_{1u}=0.2940, v_{2l}=0.5438, v_{2u}=0.7613, v_{3l}=0.0879, v_{3u}=0.1622\,.$$

The 'the characteristic vector' of R is now given below:

$$V'=(v_{1l}v_{1u}v_{2l}v_{2u}v_{3l}v_{3u})^T=(0.15080.29400.54380.76130.08790.1622)^T$$

Step 3: The priority vector of R is constructed as follows:

$$((0.1508,0.7060,0.1432)(0.5438,0.2387,0.2175)(0.0879,0.8378,0.0743))^T$$

Step 4: The scores of the vectors $(0.1508,0.7060,0.1432)$, $(0.5438,0.2387,0.2175)$ and $(0.0879,0.8378,0.0743)$ are

$$\Delta_{x_1}(0.1508,0.7060,0.1432)=0.1508-0.7060=-0.5552\,,$$
$$\Delta_{x_2}(0.5438,0.2387,0.2175)=0.5438-0.2387=0.3051\,,$$

and

$$\Delta_{x_3}(0.0879,0.8378,0.0743)=0.5438-0.2387=-0.7499$$

Because $\Delta_{x_2}>\Delta_{x_1}>\Delta_{x_3}$, the optimal ranking order of the decision alternatives is given by $x_2 \succ x_1 \succ x_3$.

The goal programming optimization model:

Step 1: Construct the goal programming optimization model as follows:

$$min\ J_1=\varepsilon_{12}^{+}+\varepsilon_{12}^{-}+\varepsilon_{13}^{+}+\varepsilon_{13}^{-}+\varepsilon_{21}^{+}+\varepsilon_{21}^{-}+\varepsilon_{23}^{+}+\varepsilon_{23}^{-}+\varepsilon_{31}^{+}+\varepsilon_{31}^{-}+\varepsilon_{32}^{+}+\varepsilon_{32}^{-}$$
$$+\gamma_{12}^{+}+\gamma_{12}^{-}+\gamma_{13}^{+}+\gamma_{13}^{-}+\gamma_{23}^{+}+\gamma_{23}^{-}+\gamma_{21}^{+}+\gamma_{21}^{-}+\gamma_{31}^{+}+\gamma_{31}^{-}+\gamma_{32}^{+}+\gamma_{32}^{-}$$

$$s.t.\begin{cases} v_{1l}-3^{-1.5}v_{2u}-\varepsilon_{12}^{+}+\varepsilon_{12}^{-}=0, v_{1l}-3^{0.5}v_{3u}-\varepsilon_{13}^{+}+\varepsilon_{13}^{-}=0, v_{2l}-3^{1.0}v_{3u}-\varepsilon_{23}^{+}+\varepsilon_{23}^{-}=0,\\ v_{2l}-3^{0.5}v_{1u}-\varepsilon_{21}^{+}+\varepsilon_{21}^{-}=0, v_{3l}-3^{-0.5}v_{1u}-\varepsilon_{31}^{+}+\varepsilon_{31}^{-}=0, v_{3l}-3^{-2.0}v_{2u}-\varepsilon_{32}^{+}+\varepsilon_{32}^{-}=0,\\ v_{1u}-3^{-0.5}v_{2l}-\gamma_{12}^{+}+\gamma_{12}^{-}=0, v_{1u}-3^{0.5}v_{3l}-\gamma_{13}^{+}+\gamma_{13}^{-}=0, v_{2u}-3^{2.0}v_{3l}-\gamma_{23}^{+}+\gamma_{23}^{-}=0,\\ v_{2u}-3^{1.5}v_{1l}-\gamma_{21}^{+}+\gamma_{21}^{-}=0, v_{3u}-3^{-0.5}v_{1l}-\gamma_{31}^{+}+\gamma_{31}^{-}=0, v_{3u}-3^{-1.0}v_{2l}-\gamma_{32}^{+}+\gamma_{32}^{-}=0,\\ v_{1l}+v_{2u}+v_{3u}\geq 1, v_{2l}+v_{1u}+v_{3u}\geq 1, v_{3l}+v_{1u}+v_{2u}\geq 1,\\ v_{1u}+v_{2l}+v_{3l}\leq 1, v_{2u}+v_{1l}+v_{3l}\leq 1, v_{3u}+v_{1l}+v_{2l}\leq 1,\\ v_{1u}-v_{1l}\geq 0, v_{2u}-v_{2l}\geq 0, v_{3u}-v_{3l}\geq 0,\\ v_{il}\geq 0, v_{iu}\geq 0, \varepsilon_{ij}^{+}\geq 0, \varepsilon_{ij}^{-}\geq 0, \gamma_{ij}^{+}\geq 0, \gamma_{ij}^{-}\geq 0, i,j\in N. \end{cases} \quad (8.96)$$

Step 2: By utilizing 'Matlab Optimization Toolbox', we have the solution to Model (8.90) as follows:

$$v_{1l}=0.1874, v_{1u}=0.2851, v_{2l}=0.5503, v_{2u}=0.6806, v_{3l}=0.1320, v_{3u}=0.1646\,.$$

The 'the characteristic vector' of R is now constructed as follows:

$$V'=(v_{1l}v_{1u}v_{2l}v_{2u}v_{3l}v_{3u})^T=(0.18740.28510.55030.68060.13200.1646)^T\,.$$

Step 3: The priority vector of R is constructed as

$$((0.1874,0.7149,0.0977)(0.5503,0.3194,0.1303)(0.1320,0.8354,0.0326))^T\,.$$

Step 4: Utilize the method of comparing intuitionistic fuzzy numbers, the optimal ranking order of the decision alternatives is also $x_2\succ x_1\succ x_3$.

Example 8.8. Suppose that the IFPR as presented by the ith decision maker based on the set $X=\{x_1,x_2,x_3\}$ of decision alternatives is R_i, $i=1,2,3$, and the weights for the individual decision makers are respectively 0.3, 0.4 and 0.3. Here, the IFPRs R_i, $i=1,2,3$ are as follows:

$$\boldsymbol{R}_1=\begin{pmatrix} (0.5,0.5,0) & (0.1,0.6,0.3) & (0.6,0.3,0.1)\\ (0.6,0.1,0.3) & (0.5,0.5,0) & (0.8,0.2,0)\\ (0.3,0.6,0.1) & (0.2,0.8,0) & (0.5,0.5,0) \end{pmatrix}$$

$$R_2=\begin{pmatrix} (0.5,0.5,0) & (0.3,0.7,0) & (0.7,0.2,0.1) \\ (0.7,0.3,0) & (0.5,0.5,0) & (0.6,0.2,0.2) \\ (0.2,0.7,0.1) & (0.2,0.6,0.2) & (0.5,0.5,0) \end{pmatrix}$$

$$R_3=\begin{pmatrix} (0.5,0.5,0) & (0.3,0.6,0.1) & (0.9,0.1,0) \\ (0.6,0.3,0.1) & (0.5,0.5,0) & (0.7,0.2,0.1) \\ (0.1,0.9,0) & (0.2,0.7,0.1) & (0.5,0.5,0) \end{pmatrix}$$

The collective IFPR R of $R_i, i=1,2,3$, is now constructed as follows:

$$R=\begin{pmatrix} (0.5,0.5,0) & (0.24,0.64,0.12) & (0.73,0.2,0.07) \\ (0.64,0.24,0.12) & (0.5,0.5,0) & (0.69,0.2,0.11) \\ (0.2,0.73,0.07) & (0.2,0.69,0.11) & (0.5,0.5,0) \end{pmatrix}$$

The least squares optimization model:

By utilizing the least squares optimization model (8.93), we obtain 'the characteristic vector' of R as follows:

$$V' = (v_{1l}v_{1u}v_{2l}v_{2u}v_{3l}v_{3u})^T = (0.05470.19640.76930.91110.02050.0342)^T$$

The priority vector of R is

$$((0.0547,0.8036,0.1417)(0.7693,0.0889,0.1418)(0.0205,0.9658,0.0137))^T$$

So, the optimal ranking order of the decision alternatives is given by $x_2 \succ x_1 \succ x_3$.

The goal programming optimization model:

By utilizing the goal programming optimal model (8.94), we obtain 'the characteristic vector' of R as well.

$$V' = (v_{1l}v_{1u}v_{2l}v_{2u}v_{3l}v_{3u})^T = (0.11810.21980.70180.85290.02900.0785)^T$$

The priority vector of R is

$$((0.1181,0.7802,0.1017)(0.7018,0.1471,0.1511)(0.0290,0.9215,0.0495))^T$$

The optimal ranking order of the decision alternatives is also $x_2 \succ x_1 \succ x_3$.

Example 8.9. Consider the set $X=\{x_1,x_2,x_3\}$ of decision alternatives, assuming that the IFPR $\boldsymbol{R}$ given by the decision maker is as follows.

$$\boldsymbol{R}=\begin{pmatrix} (0.5,0.5,0) & (0.1845,0.5,0.3155) & (0.0830,0.6309,0.2861) \\ (0.5,0.1845,0.3155) & (0.5,0.5,0) & (0.2675,0.5,0.2325) \\ (0.6309,0.0830,0.2861) & (0.5,0.2675,0.2325) & (0.5,0.5,0) \end{pmatrix}$$

The least squares optimization model:

By utilizing the least squares optimization model (8.84), we obtain 'the characteristic vector' of $\boldsymbol{R}$ as follows:

$$\boldsymbol{V}^{'}=(v_{1l}v_{1u}v_{2l}v_{2u}v_{3l}v_{3u})^{T}=(0.07210.19890.19870.40840.40850.7126)^{T}$$

The priority vector of $\boldsymbol{R}$ is

$$((0.0721,0.8011,0.1268)(0.1987,0.5916,0.2097)(0.4085,0.2874,0.3041))^{T}$$

So, the optimal ranking order of the decision alternatives is given by $x_3 \succ x_2 \succ x_1$.

The goal programming optimization model:

By utilizing the goal programming optimal model (8.86), we obtain 'the characteristic vector' V' of $\boldsymbol{R}$ as well, where:

$$\boldsymbol{V}^{'}=(v_{1l}v_{1u}v_{2l}v_{2u}v_{3l}v_{3u})^{T}=(0.07110.19590.19590.40210.40210.7024)^{T}$$

The priority vector of $\boldsymbol{R}$ is

$$((0.0711,0.8041,0.1248)(0.1959,0.5979,0.2062)(0.4021,0.2976,0.3003))^{T}$$

So, the optimal ranking order of the decision alternatives is $x_3 \succ x_2 \succ x_1$.

In fact, it is easy to verify that R is additively consistent by using Eq. (8.62).

Example 8.10. Suppose that the IFPR presented by the *ith* decision maker based on the set $X=\{x_1,x_2,x_3\}$ of decision alternatives is R_i, $i=1,2,3$, and the weights for the individual decision makers are respectively 0.2, 0.3 and 0.5. In particular, the IFPR R_i, $i=1,2,3$, are given as follows:

$$\boldsymbol{R}_1=\begin{pmatrix} (0.5,0.5,0) & (0.1845,0.5,0.3155) & (0.0830,0.6309,0.2861) \\ (0.5,0.1845,0.3155) & (0.5,0.5,0) & (0.2675,0.5,0.2325) \\ (0.6309,0.0830,0.2861) & (0.5,0.2675,0.2325) & (0.5,0.5,0) \end{pmatrix}$$

$$R_2=\begin{pmatrix} (0.5,0.5,0) & (0.3984,0.2675,0.3341) & (0.5,0.0830,0.4170) \\ (0.2675,0.3984,0.3341) & (0.5,0.5,0) & (0.3691,0.0830,0.5479) \\ (0.0830,0.5,0.4170) & (0.0830,0.3691,0.5479) & (0.5,0.5,0) \end{pmatrix}$$

$$R_3=\begin{pmatrix} (0.5,0.5,0) & (0.6016,0.1144,0.2840) & (0.4170,0.2453,0.3377) \\ (0.1144,0.6016,0.2840) & (0.5,0.5,0) & (0.1845,0.5000,0.3155) \\ (0.2453,0.4170,0.3377) & (0.5000,0.1845,0.3155) & (0.5,0.5,0) \end{pmatrix}$$

The collective IFPR R of $R_i, i=1,2,3$ is now constructed as follows:

$$R=\begin{pmatrix} (0.5,0.5,0) & (0.4572,0.2375,0.3053) & (0.3751,0.2737,0.3512) \\ (0.2375,0.4572,0.3053) & (0.5,0.5,0) & (0.2565,0.3749,0.3686) \\ (0.2737,0.3751,0.3512) & (0.3749,0.2565,0.36861) & (0.5,0.5,0) \end{pmatrix}$$

The least squares optimization model:

By utilizing the least squares optimization model (8.84), we obtain 'the characteristic vector' of $\boldsymbol{R}$ as follows:

$$V' = (v_{1l} v_{1u} v_{2l} v_{2u} v_{3l} v_{3u})^T = (0.27170.59910.14160.34370.17280.5396)^T$$

The priority vector of $\boldsymbol{R}$ is

$$((0.2717,0.4009,0.3274)(0.1416,0.6563,0.2021)(0.1728,0.4604,0.3668))^T$$

So, the optimal ranking order of the decision alternatives is given by $x_1 \succ x_3 \succ x_2$.

By utilizing the least squares optimization model (8.84), we obtain 'the characteristic vector' of R_i, $i=1,2,3$, as follows:

$$\boldsymbol{V}_1' = (v_{1l}^1\ v_{1u}^1\ v_{2l}^1\ v_{2u}^1\ v_{3l}^1\ v_{3u}^1)^T = (0.0721\ 0.1989\ 0.1987\ 0.4084\ 0.4085\ 0.7126)^T,$$

$$\boldsymbol{V}_2' = (v_{1l}^2\ v_{1u}^2\ v_{2l}^2\ v_{2u}^2\ v_{3l}^2\ v_{3u}^2)^T = (0.3138\ 0.5483\ 0.1529\ 0.5483\ 0.0555\ 0.3138)^T,$$

$$\boldsymbol{V}_3' = (v_{1l}^3\ v_{1u}^3\ v_{2l}^3\ v_{2u}^3\ v_{3l}^3\ v_{3u}^3)^T = (0.3177\ 0.7313\ 0.0881\ 0.1816\ 0.1806\ 0.5007)^T.$$

The goal programming optimization model:

By utilizing the goal programming optimal model (8.86), we obtain the following 'characteristic vector' of R as well.

$$V' = (v_{1l} v_{1u} v_{2l} v_{2u} v_{3l} v_{3u})^T = (0.28510.62860.14860.36060.18140.5662)^T$$

The priority vector of R is

$$((0.2851,0.3714,0.3435)(0.1486,0.6394,0.2120)(0.1814,0.4338,0.3848))^T$$

So, the optimal ranking order of the decision alternatives is $x_1 \succ x_3 \succ x_2$.

By utilizing the goal programming optimal model (8.86), we obtain 'the characteristic vector' of R_i, $i=1,2,3$, as follows:

$$V_1^{'} = (v_{1l}^1 v_{1u}^1 v_{2l}^1 v_{2u}^1 v_{3l}^1 v_{3u}^1)^T = (0.07110.19590.19590.40210.40210.7024)^T,$$
$$V_2^{'} = (v_{1l}^2 v_{1u}^2 v_{2l}^2 v_{2u}^2 v_{3l}^2 v_{3u}^2)^T = (0.34200.59750.16660.59750.06050.3420)^T,$$
$$V_3^{'} = (v_{1l}^3 v_{1u}^3 v_{2l}^3 v_{2u}^3 v_{3l}^3 v_{3u}^3)^T = (0.31760.73070.08840.18150.18090.5009)^T.$$

In fact, according to Eq. (8.68), $R, R_i, i=1,2,3$, are all additively consistent; and that the following equations

$$v_{il}/v_{ju} = (v_{il}^1/v_{ju}^1)^{0.2}(v_{il}^2/v_{ju}^2)^{0.3}(v_{il}^3/v_{ju}^3)^{0.5}; v_{iu}/v_{jl} = (v_{iu}^1/v_{jl}^1)^{0.2}(v_{iu}^2/v_{jl}^2)^{0.3}(v_{iu}^3/v_{jl}^3)^{0.5}$$

hold true for all $i, j = 1,2,3, i \neq j$.

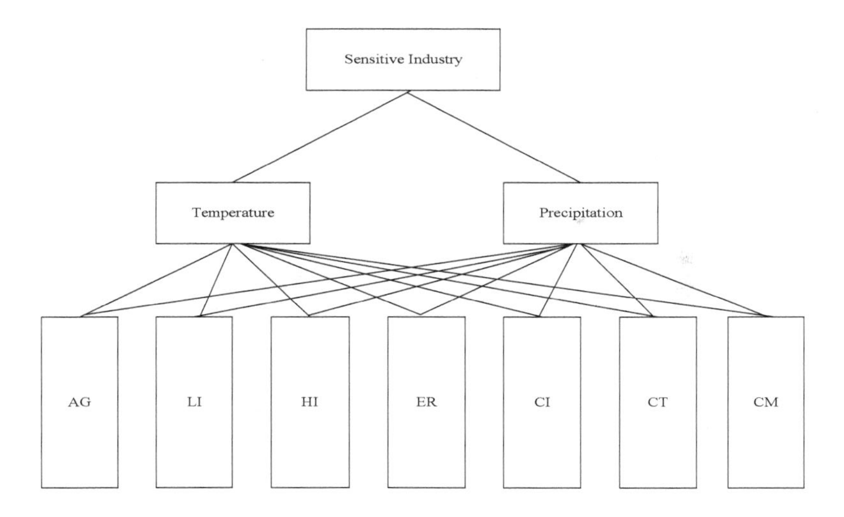

Fig. 8.5. The hierarchical structure of the MA's study

8.7.6 *IFPR Models Applied to Select Meteorologically-Sensitive Industries*

Benefit analysis and assessment of industrial meteorological service is a new endeavor of Chinese Meteorological Administration (MA). In the process of assessment we need to know the exact relationship between short-term change in meteorological conditions (temperature and precipitation) and the industrial economy, and then determine the industries that are highly sensitive to the meteorological change so that the industrial meteorological service can be evaluated. MA formed a group of experts that is composed of meteorologists, industry experts and economists to evaluate and compare the meteorological sensitivity of seven industries, including agriculture (AG), light industry (LI), heavy industry (HI), energy (ER), construction industry (CI), communications and transportation (CT), commerce (CM). The result of study is an intuitionistic fuzzy judgment matrix. The hierarchic structure of the study is given in Figure 8.5.

By using the method of pairwise comparison, the IFPR for the seven industries with respect to temperature and the IFPR of the seven industries with respect to precipitation are presented respectively in Tables 8.1 and 8.2.

Table 8.1. The IFPR for the seven industries with respect to temperature

	AG	LI	HI	ER	CI	CT	CM	Priority
AG	$(\frac{1}{2},\frac{1}{2},0)$	$(\frac{5}{8},\frac{2}{8},\frac{1}{8})$	$(\frac{7}{8},0,\frac{1}{8})$	$(\frac{6}{8},\frac{1}{8},\frac{1}{8})$	$(\frac{5}{8},\frac{1}{8},\frac{2}{8})$	$(\frac{4}{8},\frac{3}{8},\frac{1}{8})$	$(\frac{2}{8},\frac{5}{8},\frac{1}{8})$	(0.1456,0.8004,0.0540)
LI	$(\frac{2}{8},\frac{5}{8},\frac{1}{8})$	$(\frac{1}{2},\frac{1}{2},0)$	$(\frac{5}{8},\frac{2}{8},\frac{1}{8})$	$(\frac{5}{8},\frac{2}{8},\frac{1}{8})$	$(\frac{3}{8},\frac{4}{8},\frac{1}{8})$	$(\frac{2}{8},\frac{4}{8},\frac{2}{8})$	$(\frac{2}{8},\frac{6}{8},0)$	(0.0837,0.8774,0.0389)
HI	$(0,\frac{7}{8},\frac{1}{8})$	$(\frac{2}{8},\frac{5}{8},\frac{1}{8})$	$(\frac{1}{2},\frac{1}{2},0)$	$(\frac{4}{8},\frac{3}{8},\frac{1}{8})$	$(\frac{2}{8},\frac{5}{8},\frac{1}{8})$	$(\frac{2}{8},\frac{6}{8},0)$	$(0,\frac{7}{8},\frac{1}{8})$	(0.0581,0.9321,0.0098)
ER	$(\frac{1}{8},\frac{6}{8},\frac{1}{8})$	$(\frac{2}{8},\frac{5}{8},\frac{1}{8})$	$(\frac{3}{8},\frac{4}{8},\frac{1}{8})$	$(\frac{1}{2},\frac{1}{2},0)$	$(\frac{3}{8},\frac{3}{8},\frac{2}{8})$	$(\frac{2}{8},\frac{6}{8},0)$	$(0,\frac{7}{8},\frac{1}{8})$	(0.0708,0.9259,0.0033)
CI	$(\frac{1}{8},\frac{5}{8},\frac{2}{8})$	$(\frac{4}{8},\frac{3}{8},\frac{1}{8})$	$(\frac{5}{8},\frac{2}{8},\frac{1}{8})$	$(\frac{3}{8},\frac{3}{8},\frac{2}{8})$	$(\frac{1}{2},\frac{1}{2},0)$	$(\frac{2}{8},\frac{6}{8},0)$	$(\frac{1}{8},\frac{6}{8},\frac{1}{8})$	(0.0741,0.9259,0)
CT	$(\frac{3}{8},\frac{4}{8},\frac{1}{8})$	$(\frac{4}{8},\frac{2}{8},\frac{2}{8})$	$(\frac{6}{8},\frac{2}{8},0)$	$(\frac{6}{8},\frac{2}{8},0)$	$(\frac{6}{8},\frac{2}{8},0)$	$(\frac{1}{2},\frac{1}{2},0)$	$(\frac{1}{8},\frac{6}{8},\frac{1}{8})$	(0.0978,0.8641,0.0381)
CM	$(\frac{5}{8},\frac{2}{8},\frac{1}{8})$	$(\frac{6}{8},\frac{2}{8},0)$	$(\frac{7}{8},0,\frac{1}{8})$	$(\frac{7}{8},0,\frac{1}{8})$	$(\frac{6}{8},\frac{1}{8},\frac{1}{8})$	$(\frac{6}{8},\frac{1}{8},\frac{1}{8})$	$(\frac{1}{2},\frac{1}{2},0)$	(0.3418,0.6201, 0.0381)

Table 8.2. The IFPR of the seven industries with respect to precipitation

	AG	LI	HI	ER	CI	CT	CM	Priority
AG	$(\frac{1}{2},\frac{1}{2},0)$	$(\frac{6}{8},\frac{1}{8},\frac{1}{8})$	$(\frac{7}{8},0,\frac{1}{8})$	$(\frac{7}{8},\frac{1}{8},0)$	$(\frac{5}{8},\frac{2}{8},\frac{1}{8})$	$(\frac{5}{8},\frac{1}{8},\frac{2}{8})$	$(\frac{4}{8},\frac{3}{8},\frac{1}{8})$	(0.2575,0.6650,0.0775)
LI	$(\frac{1}{8},\frac{6}{8},\frac{1}{8})$	$(\frac{1}{2},\frac{1}{2},0)$	$(\frac{6}{8},\frac{1}{8},\frac{1}{8})$	$(\frac{5}{8},\frac{2}{8},\frac{1}{8})$	$(\frac{3}{8},\frac{3}{8},\frac{2}{8})$	$(\frac{3}{8},\frac{2}{8},\frac{3}{8})$	$(\frac{2}{8},\frac{5}{8},\frac{1}{8})$	(0.0802,0.8840,0.0358)
HI	$(0,\frac{7}{8},\frac{1}{8})$	$(\frac{1}{8},\frac{6}{8},\frac{1}{8})$	$(\frac{1}{2},\frac{1}{2},0)$	$(\frac{4}{8},\frac{2}{8},\frac{2}{8})$	$(\frac{3}{8},\frac{4}{8},\frac{1}{8})$	$(\frac{1}{8},\frac{6}{8},\frac{1}{8})$	$(0,\frac{7}{8},\frac{1}{8})$	(0.0643,0.9289,0.0068)
ER	$(\frac{1}{8},\frac{7}{8},0)$	$(\frac{2}{8},\frac{5}{8},\frac{1}{8})$	$(\frac{4}{8},\frac{2}{8},\frac{1}{8})$	$(\frac{1}{2},\frac{1}{2},0)$	$(\frac{3}{8},\frac{2}{8},\frac{3}{8})$	$(\frac{2}{8},\frac{5}{8},\frac{1}{8})$	$(\frac{1}{8},\frac{5}{8},\frac{2}{8})$	(0.0391,0.9164,0.0445)
CI	$(\frac{2}{8},\frac{5}{8},\frac{1}{8})$	$(\frac{3}{8},\frac{3}{8},\frac{2}{8})$	$(\frac{4}{8},\frac{3}{8},\frac{1}{8})$	$(\frac{2}{8},\frac{3}{8},\frac{3}{8})$	$(\frac{1}{2},\frac{1}{2},0)$	$(\frac{3}{8},\frac{4}{8},\frac{1}{8})$	$(\frac{2}{8},\frac{6}{8},0)$	(0.0720,0.8960,0.0320)
CT	$(\frac{1}{8},\frac{5}{8},\frac{2}{8})$	$(\frac{2}{8},\frac{3}{8},\frac{3}{8})$	$(\frac{6}{8},\frac{1}{8},\frac{1}{8})$	$(\frac{5}{8},\frac{2}{8},\frac{1}{8})$	$(\frac{4}{8},\frac{3}{8},\frac{1}{8})$	$(\frac{1}{2},\frac{1}{2},0)$	$(\frac{3}{8},\frac{4}{8},\frac{1}{8})$	(0.0802,0.8196,0.1002)
CM	$(\frac{3}{8},\frac{4}{8},\frac{1}{8})$	$(\frac{5}{8},\frac{2}{8},\frac{1}{8})$	$(\frac{7}{8},0,\frac{1}{8})$	$(\frac{5}{8},\frac{1}{8},\frac{2}{8})$	$(\frac{6}{8},\frac{2}{8},0)$	$(\frac{4}{8},\frac{3}{8},\frac{1}{8})$	$(\frac{1}{2},\frac{1}{2},0)$	(0.2101,0.7899,0)

By utilizing the goal programming optimization method (respectively, we can also use the least squares optimization method), we obtain respectively the priority of the seven industries in terms of the sensitivity on temperature, the last column of

Table 8.1, and the priority the seven industries in terms of the sensitivity on precipitation, the last column of Table 8.2. Therefore, we can see that the ranking order of the seven industries in terms of the sensitivity on temperature is $CM \succ AG \succ CT \succ LI \succ CI \succ ER \succ HI$, and the ranking order of the seven industries in terms of the sensitivity on precipitation is $AG \succ CM \succ CT \succ LI \succ CI \succ HI \succ ER$.

The expert group views the importance of temperature and that of precipitation respectively as 0.6 and 0.4. Thus, we can produce the global priority of the seven industries in terms of their meteorological sensitivity as follows:

$$(0.1904, 0.7462, 0.0634)\ (0.0823, 0.8800, 0.0377)\ (0.0606, 0.9308, 0.0086)\ (0.0581, 0.9221, 0.0198)$$
$$(0.0733, 0.9139, 0.0128)\ (0.0908, 0.8463, 0.0629)\ (0.2891, 0.6880, 0.0229).$$

Therefore, the ranking order of the seven industries in terms of their meteorological sensitivity are $CM \succ AG \succ CT \succ LI \succ CI \succ HI \succ ER$.

8.8 Optimization Models of Incomplete IFPRs

The concept of incomplete PRs has been defined in Chapter 2. In this section, we discuss the method of prioritizing incomplete IFPRs based on the multiplicative consistency. For more relevant details, please consult with (Gong, Li and Yao, 2010).

8.8.1 Optimization Models of Incomplete IFPRs

Let $X = \{x_1, x_2, \ldots, x_n\}$ be a set of decision alternatives and $d = \{d_1, d_2, \ldots, d_m\}$ a set of decision makers. The preferences of the decision makers on the set X are described by the following IFPRs:

$$\boldsymbol{R} = (\tilde{r}_{ijs})_{n\times n} = =$$

$$\begin{pmatrix} (0.5, 0.5, 0) & \left\{\begin{matrix} (\mu_{121}, \nu_{121}, \pi_{121}) \\ \vdots \\ (\mu_{12\delta_{12}}, \nu_{12\delta_{12}}, \pi_{12\delta_{12}}) \end{matrix}\right\} & \cdots & \left\{\begin{matrix} (\mu_{1n1}, \nu_{121}, \pi_{1n1}) \\ \vdots \\ (\mu_{1n\delta_{1n}}, \nu_{1n\delta_{1n}}, \pi_{1n\delta_{1n}}) \end{matrix}\right\} \\ \left\{\begin{matrix} (\mu_{211}, \nu_{211}, \pi_{211}) \\ \vdots \\ (\mu_{21\delta_{21}}, \nu_{21\delta_{21}}, \pi_{21\delta_{21}}) \end{matrix}\right\} & (0.5, 0.5, 0) & \cdots & \left\{\begin{matrix} (\mu_{2n1}, \nu_{2n1}, \pi_{2n1}) \\ \vdots \\ (\mu_{2n\delta_{2n}}, \nu_{2n\delta_{2n}}, \pi_{2n\delta_{2n}}) \end{matrix}\right\} \\ \vdots & \vdots & \cdots & \vdots \\ \left\{\begin{matrix} (\mu_{n11}, \nu_{n11}, \pi_{n11}) \\ \vdots \\ (\mu_{n1\delta_{n1}}, \nu_{n1\delta_{n1}}, \pi_{n1\delta_{n1}}) \end{matrix}\right\} & \left\{\begin{matrix} (\mu_{n21}, \nu_{n21}, \pi_{n21}) \\ \vdots \\ (\mu_{n2\delta_{n2}}, \nu_{n2\delta_{n2}}, \pi_{n2\delta_{n2}}) \end{matrix}\right\} & \cdots & (0.5, 0.5, 0) \end{pmatrix}$$

where $\tilde{r}_{ijs} = (\mu_{ijs}, v_{ijs}, \pi_{ijs})$ stands for the element of the IFPR $\tilde{R}$ satisfying the properties: $\mu_{ijs} = v_{jis}, v_{ijs} = \mu_{jis}, \pi_{ijs} = \pi_{jis}, \mu_{ijs} + v_{ijs} + \pi_{ijs} = 1 \forall i, j \in N, i \neq j, s = 1, 2, \ldots, \delta_{ij}$, and δ_{ij}, $0 \leq \delta_{ij} \leq m$, represents the number of decision makers that estimate the preference degree for the decision alternative x_i to be over the alternative x_j. It is clear that $\delta_{ij} = \delta_{ji}$. If there exist $i_0, j_0 \in N$ such that $0 < \delta_{i_0 j_0} < m$, then $m - \delta_{i_0 j_0}$ decision makers did not estimate the preference degree between the alternatives x_{i_0} and x_{j_0}; if there exists $i_0, j_0 \in N$ such that $\delta_{i_0 j_0} = 0$, then no decision maker estimated the preference degree between the alternatives x_{i_0} and x_{j_0}. In this case, we denote $\tilde{r}_{i_0 j_0 s} = -$. That means that the element $\tilde{r}_{i_0 j_0 s}$ in $\tilde{R}$ is absent and $\tilde{R}$ is an incomplete IFPR. If for all $i, j \in N$, $\delta_{ij} = m$, then it means that all the decision makers decide on the preference between the alternatives x_i and x_j, and $\tilde{R}$ is a complete IFPR.

Suppose that all decision makers have the same degree of preference for the alternative x_i to be over x_j, $\forall i, j \in N$. For the given multiplicative consistent IFPR $\tilde{R} = (\mu_{ijs}, v_{ijs}, \pi_{ijs})_{n \times n}$, there must exist a priority vector

$$\zeta = ((\omega_{1l}, 1 - \omega_{1u}, \omega_{1u} - \omega_{1l}) \ldots (\omega_{il}, 1 - \omega_{iu}, \omega_{iu} - \omega_{il}) \ldots (\omega_{nl}, 1 - \omega_{nu}, \omega_{nu} - \omega_{nl}))^T$$

such that

$$\mu_{ijs} = \frac{\omega_{il}}{\omega_{il} + \omega_{ju}} \tag{8.97}$$

$$v_{ijs} = 1 - \frac{\omega_{iu}}{\omega_{jl} + \omega_{iu}} = \frac{\omega_{jl}}{\omega_{jl} + \omega_{iu}} \tag{8.98}$$

where $0 < \omega_{il} \leq \omega_{iu} < 1, i, j \in N$. So, Eqs. (8.97) and (8.98) are equivalent to the following:

$$\mu_{ijs} \omega_{ju} - (1 - \mu_{ijs}) \omega_{il} = 0 \tag{8.99}$$

$$v_{ijs} \omega_{iu} - (1 - v_{ijs}) \omega_{jl} = 0 \tag{8.100}$$

Eqs. (8.99) and (8.100) actually stand for the ideal cases. In reality, it is hard for a decision maker to be consistent, and different decision makers may present different judgments. In consequence, Eqs. (8.99) and (8.100) may not hold true in general. For such a situation, consider the following deviation functions:

$$\varepsilon_{ijs} = [\mu_{ijs} \omega_{ju} - (1 - \mu_{ijs}) \omega_{il}]^2 \tag{8.101}$$

$$\gamma_{ijs} = [v_{ijs} \omega_{iu} - (1 - v_{ijs}) \omega_{jl}]^2 \tag{8.102}$$

It is clear that small deviation functions represent better judgment consistency. In order to obtain the optimal priority vector of inconsistent IFPRs, we introduce a least squares optimal model as follows:

$$minJ = \sum_{i=1}^{n} \sum_{j=1, j\neq i}^{n} \sum_{s=1}^{\delta_{ij}} [\mu_{ijs}\omega_{ju} - (1-\mu_{ijs})\omega_{il}]^2 + [\nu_{ijs}\omega_{iu} - (1-\nu_{ijs})\omega_{jl}]^2$$

$$s.t.\begin{cases} \omega_{il} + \sum_{j=1, j\neq i}^{n} \omega_{ju} \geq 1, & i \in N; \\ \omega_{iu} + \sum_{j=1, j\neq i}^{n} \omega_{jl} \leq 1, & i \in N; \\ \omega_{iu} - \omega_{il} \geq 0, & i \in N; \\ \omega_{iu} \geq 0, \omega_{il} \geq 0, & i \in N \end{cases} \tag{8.103}$$

The constraints $\omega_{il} + \sum_{j=1, j\neq i}^{n} \omega_{ju} \geq 1$ and $\omega_{iu} + \sum_{j=1, j\neq i}^{n} \omega_{jl} \leq 1$ are used for the normalization of the interval vector ω (Sugihara, Ishii and Tanaka, 2004).

In model (8.103), if $\delta_{ij} = m$, $\forall i, j \in N$, we then obtain a collective priority model of the IFPRs as presented by the m decision makers with complete information. If $\delta_{ij} = 1$, $\forall i, j \in N$, we then obtain a priority model of the individual IFPRs with complete information. If there exist $i_0, j_0 \in N$ such that $0 < \delta_{i_0 j_0} < m$, then we obtain a collective priority model of the IFPRs presented by the m decision makers with incomplete information.

8.8.2 A Numerical Example

Example 8.11. Suppose that there are three decision makers who provide the following incomplete IFPRs $\{\tilde{R}_1, \tilde{R}_2, \tilde{R}_3\}$ on a set $X = \{x_1, x_2, x_3, x_4\}$ of four decision alternatives.

$$\boldsymbol{R}_1 = \begin{pmatrix} (0.5,0.5,0) & (0.1,0.6,0.3) & - & - \\ (0.6,0.1,0.3) & (0.5,0.5,0) & (0.8,0.2,0) & - \\ - & (0.2,0.8,0) & (0.5,0.5,0) & (0.8,0.1,0.1) \\ - & - & (0.1,0.8,0.1) & (0.5,0.5,0) \end{pmatrix}$$

$$\boldsymbol{R}_2 = \begin{pmatrix} (0.5,0.5,0) & (0.2,0.7,0.1) & - & - \\ (0.7,0.2,0.1) & (0.5,0.5,0) & (0.7,0.1,0.2) & - \\ - & (0.1,0.7,0.2) & (0.5,0.5,0) & - \\ - & - & - & (0.5,0.5,0) \end{pmatrix}$$

$$\boldsymbol{R}_3 = \begin{pmatrix} (0.5,0.5,0) & (0.1,0.6,0.3) & (0.7,0.1,0.2) & - \\ (0.6,0.1,0.3) & (0.5,0.5,0) & - & (0.8,0.1,0.1) \\ (0.1,0.7,0.2) & - & (0.5,0.5,0) & - \\ - & (0.1,0.8,0,1) & - & (0.5,0.5,0) \end{pmatrix}$$

Step 1: By using model (8.103), we first construct the following optimization model:

$$\begin{aligned} minJ_1 = {} & (0.1\omega_{2u}-0.9\omega_{1l})^2+(0.6\omega_{1u}-0.4\omega_{2l})^2+(0.2\omega_{2u}-0.8\omega_{1l})^2+(0.7\omega_{1u}-0.3\omega_{2l})^2 \\ & +(0.1\omega_{2u}-0.9\omega_{1l})^2+(0.6\omega_{1u}-0.4\omega_{2l})^2+(0.7\omega_{3u}-0.3\omega_{1l})^2+(0.1\omega_{1u}-0.9\omega_{3l})^2 \\ & +(0.6\omega_{1u}-0.4\omega_{2l})^2+(0.1\omega_{2u}-0.9\omega_{1l})^2+(0.7\omega_{1u}-0.3\omega_{2l})^2+(0.2\omega_{2u}-0.8\omega_{1l})^2 \\ & +(0.6\omega_{1u}-0.4\omega_{2l})^2+(0.1\omega_{2u}-0.9\omega_{1l})^2+(0.8\omega_{3u}-0.2\omega_{2l})^2+(0.2\omega_{2u}-0.8\omega_{3l})^2 \\ & +(0.7\omega_{3u}-0.3\omega_{2l})^2+(0.1\omega_{2u}-0.9\omega_{3l})^2+(0.8\omega_{4u}-0.2\omega_{2l})^2+(0.1\omega_{2u}-0.9\omega_{4l})^2 \\ & +(0.1\omega_{1u}-0.9\omega_{3l})^2+(0.7\omega_{3u}-0.3\omega_{1l})^2+(0.2\omega_{2u}-0.8\omega_{3l})^2+(0.8\omega_{3u}-0.2\omega_{2l})^2 \\ & +(0.1\omega_{2u}-0.9\omega_{3l})^2+(0.7\omega_{3u}-0.3\omega_{2l})^2+(0.8\omega_{4u}-0.2\omega_{3l}))^2+(0.1\omega_{3u}-0.9\omega_{4l})^2 \\ & +(0.1\omega_{2u}-0.9\omega_{4l})^2+(0.8\omega_{4u}-0.2\omega_{2l})^2+(0.1\omega_{3u}-0.9\omega_{4l})^2+(0.8\omega_{4u}-0.2\omega_{3l})^2 \end{aligned} \tag{8.104}$$

$$s.t.\begin{cases} \omega_{1l}+\omega_{2u}+\omega_{3u}+\omega_{4u}\ge 1, \omega_{2l}+\omega_{1u}+\omega_{3u}+\omega_{4u}\ge 1, \\ \omega_{3l}+\omega_{1u}+\omega_{2u}+\omega_{4u}\ge 1, \omega_{4l}+\omega_{1u}+\omega_{2u}+\omega_{3u}\ge 1, \\ \omega_{1u}+\omega_{2l}+\omega_{3l}+\omega_{4l}\le 1, \omega_{2u}+\omega_{1l}+\omega_{3l}+\omega_{4l}\le 1, \\ \omega_{3u}+\omega_{1l}+\omega_{2l}+\omega_{4l}\le 1, \omega_{4u}+\omega_{1l}+\omega_{2l}+\omega_{3l}\le 1, \\ \omega_{1u}-\omega_{1l}\ge 0, \omega_{2u}-\omega_{2l}\ge 0, \\ \omega_{3u}-\omega_{3l}\ge 0, \omega_{4u}-\omega_{4l}\ge 0, \\ \omega_{il}\ge 0, \omega_{iu}\ge 0, i,j=1,2,3,4. \end{cases}$$

Step 2: By using 'Matlab Optimization Toolbox', we obtain the following solutions to Model (8.104):

$$\omega_{1l}=0.1132; \omega_{1u}=0.2876; \omega_{2l}=0.4791; \omega_{2u}=0.6536;$$

$$\omega_{3l}=0.0886; \omega_{3u}=0.1418; \omega_{4l}=0.0442; \omega_{4u}=0.0914$$

Step 3: The priority vector of $\tilde{R}$ is found to be

$$((0.113,0.712,0.174)(0.479,0.346,0.175)(0.089,0.858,0.053)(0.044,0.909,0.047))^T$$

Step 4: By using the method of comparing two intuitionistic fuzzy values, as studied in Section 8.2, the optimal ranking order of the decision alternatives is found to be $x_2 \succ x_1 \succ x_3 \succ x_4$.

8.9 Group Decision Making Model Based on IFPRs

Except for the methods of comparing intuitionistic fuzzy values, we are seldom concerned about any operation law of intuitionistic fuzzy values. In this section, by using some basic operation laws of intuitionistic fuzzy values (Xu and Yager, 2006; Xu,2007a,2007b), we discuss the group decision making problem based on IFPRs (Tang and Gong, 2007; Gong and Liu, 2007b).

8.9.1 Operation Laws of Intuitionistic Fuzzy Values

This section is mainly based on (Xu and Yager, 2006; Xu, 2007a, 2007b).

Let $\tilde{a}_i = [\mu_i, 1-v_i], i \in N = \{1, 2, \ldots, n\}$ be n intuitionistic fuzzy values, then the laws of operation of these intuitionistic fuzzy values are given as follows:

$$\text{Addition: } \tilde{a}_i \oplus \tilde{a}_j = [\mu_i + \mu_j - \mu_i\mu_j, 1 - v_i v_j] \tag{8.105}$$

$$\text{Scalar multiplication: } \lambda\tilde{a}_i = [1-(1-\mu_i)^\lambda, 1-v_i^\lambda] \tag{8.106}$$

$$\text{Comparison: } \tilde{a}_i \le \tilde{a}_j \Leftrightarrow \mu_i \le \mu_j, v_i \ge v_j \tag{8.107}$$

By using Eqs. (8.105) and (8.106), we can readily obtain the following:

$$\lambda(\tilde{a}_1 \oplus \cdots \oplus \tilde{a}_n) = \lambda\tilde{a}_1 \oplus \cdots \oplus \lambda\tilde{a}_n = \left[1-\prod_{i=1}^{n}(1-\mu_i)^\lambda,\ 1-\prod_{i=1}^{n}\mu_i^{\ \lambda}\right] \tag{8.108}$$

where $\lambda > 0$, and if $\lambda = \frac{1}{n}$, Eq. (8.108) is referred to as the arithmetic average of the given intuitionistic fuzzy values.

For the sake of convenience of our discussion, we also refer $(\mu_A(x), v_A(x), \pi_A(x))$ to as an intuitionistic fuzzy value in this text. So $\tilde{a}_i = [\mu_i, 1-v_i]$ can be also denoted by $\tilde{a}_i = (\mu_i, v_i, \pi_i), i \in N$, where $\pi_i = 1 - v_i - \mu_i$. Some equivalent forms of Eqs. (8.105), (8.106) and (8.107) are given below:

$$\tilde{a}_i \oplus \tilde{a}_j = (\mu_i + \mu_j - \mu_i\mu_j, v_i v_j, 1 - \mu_i - \mu_j + \mu_i\mu_j - v_i v_j) \tag{8.109}$$

$$\lambda\tilde{a}_i = (1-(1-\mu_i)^\lambda, v_i^\lambda, (1-\mu_i)^\lambda - v_i^\lambda) \tag{8.110}$$

$$\tilde{a}_i \le \tilde{a}_j \Leftrightarrow \mu_i \le \mu_j, v_i \ge v_j \tag{8.111}$$

$$\lambda(\tilde{a}_1 \oplus \cdots \oplus \tilde{a}_n) = \lambda\tilde{a}_1 \oplus \cdots \oplus \lambda\tilde{a}_n = (1-\prod_{i=1}^{n}(1-\mu_i)^\lambda,\ \prod_{i=1}^{n}\mu_i^{\ \lambda}, (1-\mu_i)^\lambda - \prod_{i=1}^{n}\mu_i^{\ \lambda}) \tag{8.112}$$

If $\lambda = \frac{1}{n}$, we also refer Eq. (8.112) to as the arithmetic average of the given intuitionistic fuzzy values.

Definition 8.21. If we consider $\tilde{a}_i = (\mu_i, v_i, \pi_i), \pi_i = 1 - v_i - \mu_i, i \in N$, as 3-dimensional vectors (In the rest of this text we also regard intuitionistic fuzzy values as vectors. By doing so it makes it easier for us to discuss relevant problems), then the cosine of the angle between the vectors $\tilde{a}_i$ and $\tilde{a}_j$

$$C(\tilde{a}_i, \tilde{a}_j) = \frac{\mu_i \mu_j + v_i v_j + \pi_i \pi_j}{(\mu_i^2 + v_i^2 + \pi_i^2)^{\frac{1}{2}} (\mu_j^2 + v_j^2 + \pi_j^2)^{\frac{1}{2}}}$$

can be defined as the degree of correlation between the intuitionistic fuzzy values $\tilde{a}_i$ and $\tilde{a}_j$.

It is obvious that $0 \le C(\widetilde{a_i}, \widetilde{a_j}) \le 1$, and the greater $C(\widetilde{a_i}, \widetilde{a_j})$ is, the larger the degree of correlation between the two values becomes. Note: We can also use the symbol $C^2(\widetilde{a_i}, \widetilde{a_j}) = C(\widetilde{a_i}, \widetilde{a_j}) \times C(\widetilde{a_i}, \widetilde{a_j})$ to denote the degree of correlation between $\tilde{a}_i$ and $\tilde{a}_j$.

If we transform the vector $\widetilde{a_i}, \widetilde{a_j}$ to unit vectors, that is

$$\widetilde{a_i}' = \frac{(\mu_i, v_i, \pi_i)}{(\mu_i^2 + v_i^2 + \pi_i^2)^{\frac{1}{2}}}; \widetilde{a_j}' = \frac{(\mu_j, v_j, \pi_j)}{(\mu_j + v_j + \pi_j)^{\frac{1}{2}}}.$$

then the equations $C(\tilde{a}_i, \tilde{a}_j) = C(\tilde{a}_i', \tilde{a}_j')$ and $C^2(\tilde{a}_i, \tilde{a}_j) = C^2(\tilde{a}_i', \tilde{a}_j')$ also hold true.

Definition 8.22. Let $D(\tilde{a}_i, \tilde{a}_j) = [(\mu_i - \mu_j)^2 + (v_i - v_j)^2 + (\pi_i - \pi_j)]^{\frac{1}{2}}$ denote the Euclidean distance between $\tilde{a}_i$ and $\tilde{a}_j$. Then $S(\tilde{a}_i, \tilde{a}_j) = 1 - \frac{1}{3} D^2(\tilde{a}_i, \tilde{a}_j)$ denotes the degree of similarity between $\tilde{a}_i$ and $\tilde{a}_j$.

Obviously, the greater $S(\tilde{a}_i, \tilde{a}_j)$ is, the larger the degree of similarity is between $\tilde{a}_i$ and $\tilde{a}_j$.

Both the degree of correlation and the degree of similarity can be regarded as measures of the degree of consistency between two arbitrarily given intuitionistic fuzzy values.

8.9.2 *Group Decision Making Model of IFPRs*

Generally speaking, due to the conflict of either opinions or interests of the individual decision makers, the aim of the group decision is to reach a compromise or a consensus. The compromise or consensus can be determined by using a standard that all the decision makers agree with. In fact, this standard can be seen as the common preference of the decision makers. In the following, we will show how to aggregate different individual preferences in the following three different formats:

- Intuitionistic fuzzy arithmetic average.
- From the point of correlation, the maximization of the consistency of the individual decision makers preferences.
- From the point of similarity, the maximization of the consistency of the individual decision makers preferences.

Let X stand for a decision-making problem, $S = \{s_1, s_2, \ldots, s_n\}$ be a set of decision alternatives, and $D = \{d_1, d_2, \ldots, d_m\}$ a set of decision makers. Under a given criterion, the intuitionistic fuzzy value as provided by the ith decision maker $d_i, i \in M = \{1, 2, \ldots, m\}$, is $R_i = (\mu_i, v_i, \pi_i)$, where

$$\mu_i + v_i + \pi_i = 1, 0 \le \mu_i \le 1, 0 \le v_i \le 1, 0 \le \pi_i \le 1 .$$

8.9.2.1 Aggregation Using Arithmetic Average

Let $R_* = f(R_1, R_2, \cdots, R_m)$ be the ultimate unanimous opinion, where $R_i, i \in M$, is the intuitionistic fuzzy estimation value of the ith decision maker. Let function f denote the arithmetic average of these m decision makers', then the result of aggregation is given as follows:

$$R_* = f(R_1, R_2, \cdots, R_m) = \frac{1}{m} R_1 \oplus \frac{1}{m} R_2 \oplus \cdots \oplus \frac{1}{m} R_m = \frac{1}{m}(R_1 \oplus R_2 \oplus \cdots \oplus R_m)$$
$$= (1 - \prod_{i=1}^{m} (1-\mu_i)^{\frac{1}{m}}, \prod_{i=1}^{m} v_i^{\frac{1}{m}}, \prod_{i=1}^{m} (1-\mu_i)^{\frac{1}{m}} - \prod_{i=1}^{m} v_i^{\frac{1}{m}}).$$

8.9.2.2 Optimal Aggregation Based on Correlation

As we have mentioned previously, the essence of each decision-making process is to reach a compromise or a consensus. Therefore, we must find an "ideal" estimation value as presented by an "ideal" decision maker: The degree of consistency between this "ideal" estimation value and the estimation value of each decision maker is larger than the degree of consistency between the estimation values of any two decision makers. This degree of consistency can be measured by using the method of correlation.

Suppose that the "ideal" estimation value $R_* = (\mu_*, v_*, \pi_*)$ is given by the "ideal" decision maker. Then the following equation holds true:

$$max \frac{1}{m} \sum_{i=1}^{m} C^2(R, R_i) = \frac{1}{m} \sum_{i=1}^{m} C^2(R_*, R_i) \tag{8.113}$$

where $R = (\mu, v, \pi)$ stands for an intuitionistic fuzzy value.

In order to theoretically produce the optimal value $R_* = (\mu_*, v_*, \pi_*)$, we first introduce the following two results.

Lemma 8.4. (Rayleigh (Horn and Johnson,1990)). Let $A \in M_n$. If $A = A^T$, then $max\ r(x) = max \frac{x^T Ax}{x^T x} = \lambda_{max}$, where λ_{max} stands for the largest eigenvalue of A. The expression $r(x) = \frac{x^T Ax}{x^T x}$ is known as Rayleigh-Ritz ratio, and $x \neq 0$ is an n-dimensional column vector.

Lemma 8.5. (Frobenius (Horn and Johnson,1990)). Let $R = (r_{ij})_{n \times n} \in M_n$ be irreducible and nonnegative, then

1) R has one and only one maximum eigenvalue λ_{max}; and

2) All the entries of the eigenvector corresponding to the eigenvalue λ_{max} must be positive, and the only difference among all the eigenvectors of R is their ratio factors.

Let $b = (b_1, b_2, b_3) = (\mu', v', \pi')$, $x_i = (\mu_i', v_i', \pi_i')$, and $b_* = (\mu_*', v_*', \pi_*')$ be respectively the unit vectors of $R = (\mu, v, \pi)$, $R_i = (\mu_i, v_i, \pi_i)$, and $R_* = (\mu_*, v_*, \pi_*)$, $i \in M$. Then Eq. (8.113) can be denoted as:

$$max \frac{1}{m} \sum_{i=1}^{m} [(\mu', v', \pi')(\mu_i', v_i', \pi_i')^T]^2 = \frac{1}{m} \sum_{i=1}^{m} [(\mu_*', v_*', \pi_*')(\mu_i', v_i', \pi_i')^T]^2 \tag{8.114}$$

Let

$$f(b) = \sum_{i=1}^{m} [(\mu', v', \pi')(\mu_i', v_i', \pi_i')^T]^2 = \sum_{i=1}^{m} (bx_i^T bx_i^T) = b(\sum_{i=1}^{m} x_i^T x_i) b^T = bX^T X b^T \tag{8.115}$$

where

$$X = \begin{pmatrix} \mu_1' & v_1' & \pi_1' \\ \mu_2' & v_2' & \pi_2' \\ \cdots & \cdots & \cdots \\ \mu_m' & v_m' & \pi_m' \end{pmatrix}$$

Obviously, in Eq. (8.115), $X^T X$ is a symmetric matrix, and $bb^T = 1$. By using Lemma 8.4, we have that $max\ f(b) = max \frac{bX^T X b^T}{bb^T} = \lambda_{max}$, where λ_{max} is the maximum eigenvalue of $X^T X$. In the following, we will show that b_* is the unique eigenvector of $X^T X$ corresponding to the eigenvalue λ_{max}.

To this end, let us construct a Lagrange function

$$g(b,\lambda) = f(b) - \lambda(bb^T - 1) \tag{8.116}$$

By letting $\frac{\partial g(b,\lambda)}{\partial b_i} = 0, i = 1,2,3$, we have

$$X^T Xb = \lambda b \tag{8.117}$$

It is ready to see that $X^T X$ is irreducible. By using Lemma 8.5, we have that $X^T X$ must have only one maximum eigenvalue λ_{max}, and the entries of the eigenvector b_* of $X^T X$ corresponding to the eigenvalue λ_{max} are all positive. Because b_* is a unit vector, b_* must be the unique vector corresponding to the eigenvalue λ_{max}. From Eq. (8.117), it follows that if we can obtain the maximum value of f, then λ_{max} is the eigenvalue of $X^T X$, and b_* the eigenvector of $X^T X$ corresponding to λ_{max}.

Therefore, we have shown the following result.

Theorem 8.15. For any $b = (b_1, b_2, b_3) = (\mu', v', \pi')$ and any

$$X = \begin{pmatrix} \mu_1' & v_1' & \pi_1' \\ \mu_2' & v_2' & \pi_2' \\ \cdots & \cdots & \cdots \\ \mu_m' & v_m' & \pi_m' \end{pmatrix},$$

we have $max\, bX^T Xb^T = b_* X^T Xb_*^T = \lambda_{max}$, where λ_{max} is the maximum eigenvalue of $X^T X$, b_* the unique positive eigenvector corresponding to λ_{max}, and $b_* b_*^T = 1$.

QED.

According to Theorem 8.15, if $b_* = (\mu_*', v_*', \pi_*')$ has been obtained, then by using both of the equations $\mu_* + v_* + \pi_* = 1$ and $(\mu_*', v_*', \pi_*') = (\mu_*, v_*, \pi_*)(\mu_*^2 + v_*^2 + \pi_*^2)^{-\frac{1}{2}}$, the value of $R_* = (\mu_*, v_*, \pi_*)$ can be readily produced.

8.9.2.3 Aggregation Based on Similarity

In each decision-making problem, the degree of consistency between the estimation value of the "ideal" decision maker and the estimation value of each actual decision maker is larger than the degree of consistency between the estimation values of any two actual decision makers. Such consistencies can be measured by using the method of similarity. That is, R_* should satisfy the following equation.

$$max\{1-\frac{1}{3m}\sum_{i=1}^{m}[(\mu-\mu_i)^2+(v-v_i)^2+(\pi-\pi_i)^2]\}$$
$$=1-\frac{1}{3m}\sum_{i=1}^{m}[(\mu_*-\mu_i)^2+(v_*-v_i)^2+(\pi_*-\pi_i)^2] \tag{8.118}$$

Theorem 8.16. The optimal solution to the nonlinear optimization model (8.118) is

$$\mu_*=\tfrac{1}{m}\sum_{i=1}^{m}\mu_i,\ v_*=\tfrac{1}{m}\sum_{i=1}^{m}v_i,\ \pi_*=\tfrac{1}{m}\sum_{i=1}^{m}\pi_i$$

Proof. This optimization problem is quite elementary, and the proof of this result can be found in many nonlinear programming books, so we omit all the relevant details. QED.

8.9.2.4 Relationship between These New Models

Suppose that the estimation values of a group decision-making, as respectively derived by using the method of intuitionistic fuzzy arithmetic average, the method of correlation and the method of similarity, are given as follows:

$$R_*^1=(\mu_*^1,v_*^1,\pi_*^1), R_*^2=(\mu_*^2,v_*^2,\pi_*^2), R_*^3=(\mu_*^3,v_*^3,\pi_*^3),$$

where $R_i, i=1,2,\ldots,m$, stand for the estimation values of the individual decision makers.

By making use of the relation between arithmetic averages and geometric averages, we have

$$1-\prod_{i=1}^{m}(1-\mu_i)^{\frac{1}{m}}\geq\frac{1}{m}\sum_{i=1}^{m}\mu_i \tag{8.119}$$

$$\prod_{i=1}^{m}v_i^{\frac{1}{m}}\leq\frac{1}{m}\sum_{i=1}^{m}v_i \tag{8.120}$$

By comparing intuitionistic fuzzy values, we have

$$R_*^1=(1-\prod_{i=1}^{m}(1-\mu_i)^{\frac{1}{m}},\prod_{i=1}^{m}v_i^{\frac{1}{m}},\prod_{i=1}^{m}(1-\mu_i)^{\frac{1}{m}}-\prod_{i=1}^{m}v_i^{\frac{1}{m}})\geq(\frac{1}{m}\sum_{i=1}^{m}\mu_i,\frac{1}{m}\sum_{i=1}^{m}v_i,\frac{1}{m}\sum_{i=1}^{m}\pi_i)=R_*^2 \tag{8.121}$$

Thus we have shown the following result.

Property 8.1. The estimation value of the group decision-making, as derived by using the method of intuitionistic fuzzy arithmetic average satisfies $R_*^1\geq R_*^2$.

Based on Eqs. (8.113) and (8.118), it is ready for us to see the properties below.

Property 8.2. The "ideal" estimation value of the group decision-making, as derived by employing the method of correlation satisfies:

$$\frac{1}{m}\sum_{i=1}^{m} C^2(R_*^2, R_i) \geq \frac{1}{m}\sum_{i=1}^{m} C^2(R_*^t, R_i), t = 1,3$$

Property 8.3. The "ideal" estimation value of the group decision-making derived by similarity method satisfies

$$1-\frac{1}{3m}\sum_{i=1}^{m}[(\mu_*^3-\mu_i)^2+(v_*^3-v_i)^2+(\pi_*^3-\pi_i)^2]$$

$$\geq 1-\frac{1}{3m}\sum_{i=1}^{m}[(\mu_*^t-\mu_i)^2+(v_*^t-v_i)^2+(\pi_*^t-\pi_i)^2], t=1,2$$

From these results, we can establish the following conclusions:

1) The consensus of a group decision-making, as derived respectively by utilizing the method of correlation and the method of similarity, is the arithmetic average of the elements in the same position of the intuitionistic fuzzy values as provided by the individual decision makers. Because the similarity model is actually a least squares optimization model, while the correlation model is a variation of the least square optimization model, the difference between these two "ideal" estimation values as derived respectively from these two methods is very small.

2) From Eqs. (8.119) – (8.121), it follows that the group preference, as derived by using the intuitionistic fuzzy arithmetic average, is more inclinable to the "positive" attitude of a certain problem, while the group preference, as derived by using either the method of correlation or the method of similarity, is more inclinable to the "negative" attitude of the problem.

8.9.3 A Numerical Example

Example 8.12. Consider a group decision-making problem with a given set $X=\{x_1,x_2,x_3\}$ of decision alternatives by using the opinions of three decision makers k_1, k_2, and k_3. Suppose that the relevant IFPRs are given as follows:

$$\boldsymbol{M}^1=\begin{pmatrix}(0.5,0.5,0) & (0.2,0.6,0.2) & (0.4,0.5,0.1)\\(0.6,0.2,0.2) & (0.5,0.5,0) & (0.6,0.3,0.1)\\(0.5,0.4,0.1) & (0.3,0.6,0.1) & (0.5,0.5,0)\end{pmatrix}$$

$$\boldsymbol{M}^2=\begin{pmatrix} (0.5,0.5,0) & (0.3,0.6,0.1) & (0.6,0.1,0.3) \\ (0.6,0.3,0.1) & (0.5,0.5,0) & (0.5,0.3,0.2) \\ (0.1,0.6,0.3) & (0.3,0.5,0.2) & (0.5,0.5,0) \end{pmatrix}$$

$$\boldsymbol{M}^3=\begin{pmatrix} (0.5,0.5,0) & (0.2,0.6,0.2) & (0.7,0.2,0.1) \\ (0.6,0.2,0.2) & (0.5,0.5,0) & (0.6,0.4,0) \\ (0.2,0.7,0.1) & (0.4,0.6,0) & (0.5,0.5,0) \end{pmatrix}$$

With every aggregated element in the same position of these three matrices obtained by respectively applying the different methods as mentioned in the previous subsection, the corresponding group judgment matrices can be obtained as follows:

A. The group decision-making matrix, as derived from the method of intuitionistic fuzzy arithmetic average, is

$$\boldsymbol{M}_*^1=\begin{pmatrix} (0.5,0.5,0) & (0.2348,0.6000,0.1652) & (0.5840,0.2154,0.2006) \\ (0.6000,0.2348,0.1652) & (0.5,0.5,0) & (0.5691,0.3302,0.1007) \\ (0.2154,0.5840,0.2006) & (0.3302,0.5691,0.1007) & (0.5,0.5,0) \end{pmatrix}$$

B. The group decision-making matrix derived through correlation method is

$$\boldsymbol{M}_*^2=\begin{pmatrix} (0.5,0.5,0) & (0.2337,0.6000,0.1663) & (0.5651,0.2670,0.1679) \\ (0.6000,0.2337,0.1663) & (0.5,0.5,0) & (0.5640,0.3308,0.1052) \\ (0.2670,0.5651,0.1679) & (0.3357,0.5696,0.0946) & (0.5,0.5,0) \end{pmatrix}$$

C. The group decision-making matrix derived through similarity method is

$$\boldsymbol{M}_*^3=\begin{pmatrix} (0.5,0.5,0) & (0.2333,0.6000,0.1667) & (0.5667,0.2667,0.1667) \\ (0.6000,0.2333,0.1667) & (0.5,0.5,0) & (0.5667,0.3333,0.1000) \\ (0.2667,0.5667,0.1667) & (0.3333,0.5667,0.1000) & (0.5,0.5,0) \end{pmatrix}$$

By calculating the arithmetic average of the elements in the same row while excluding the elements of the principal diagonal of the matrix of each group decision-making matrix, we obtain the overall degree $t_i, i=1,2,3$, of how much the decision alternative $s_i, i=1,2,3$, is preferred to all the other alternatives. In particular, For M_*^1, we have

$$t_1=(0.4358,0.3595,0.2047), t_2=(0.5848,0.2784,0.1368), t_3=(0.2751,0.5765,0.1484)$$

Thus the optimal ranking order of the decision alternatives is given by $x_2 \succ x_1 \succ x_3$.

For M_*^2, we have

$$t_1 = (0.4227, 0.4002, 0.1771), t_2 = (0.5824, 0.2780, 0.1396), t_3 = (0.3022, 0.5673, 0.1305)$$

Thus the optimal ranking order of the decision alternatives is given by $x_2 \succ x_1 \succ x_3$.

For M_*^3, we have

$$t_1 = (0.4236, 0.4000, 0.1764), t_2 = (0.5837, 0.2789, 0.1374), t_3 = (0.3008, 0.5667, 0.1325)$$

Thus the optimal ranking order of the decision alternatives is given by $x_2 \succ x_1 \succ x_3$.

8.10 Consensus Measures of Group IFPRs

In this section, we measure the degree of consensus (or correlation) of group IFPRs.

8.10.1 Consensus Measure of IFPRs

This subsection is mainly based on (Tang and Gong, 2007; Gong and Liu, 2007b).

As is known, the aim of group decision making is to reach a compromise or a consensus. In each group decision making with IFPRs, an ideal IFPR usually needs to be constructed so that the degree of consistency between this ideal IFPR and all the individual IFPRs is higher than the degree of consistency between any two other IFPRs and all the individual IFPRs. In other words, we must somehow find an ideal matrix whose entries are highly consistent with the corresponding entries of each individual IFPR used in the group decision making. Suppose that the ideal IFPR $M^* = (m_{ij}^*)_{n\times n} = (r_{ij}^*, t_{ij}^*, \pi_{ij}^*)_{n\times n}$ is presented by an ideal decision maker $*$. Let

$$Cor^{*,k_p}(i,j) = C(m_{ij}^*, m_{ij}^{k_p}) = C((r_{ij}^*, t_{ij}^*, \pi_{ij}^*), (r_{ij}^{k_p}, t_{ij}^{k_p}, \pi_{ij}^{k_p})) \tag{8.122}$$

denote the degree of consistency (correlation) between the (i,j) entry in M^* and that in the k_pth IFPR, $k_p \in \mathrm{L} = \{1, 2, \ldots, l\}$. Obviously, the bigger the $Cor^{*,k_p}(i,j)$ value is, the higher consistency (correlation) between the ideal decision maker $*$ and the decision maker k_p in terms of the degree of how much the decision alternative x_i is preferred to the alternative x_j.

Let

$$Cor(i,j) = \frac{1}{l}\sum_{p=1}^{l} Cor^{*,k_p}(i,j) \tag{8.123}$$

denote the degree of consistency (correlation) between the (i,j) entry in M^* and those in all the individual IFPRs. Obviously, the bigger the $Cor(i,j)$ value is, the higher consistency (correlation) between the ideal decision maker $*$ and all the individual decision makers in terms of the degree of how much the decision alternative x_i is preferred to the alternative x_j.

Let

$$Cor^{*,k_p} = \frac{1}{C_n^2}\sum_{i=1}^{n-1}\sum_{j=i+1}^{n} Cor^{*,k_p}(i,j) \tag{8.124}$$

denote the degree of consistency (correlation) between the ideal decision maker and the k_pth decision maker, $k_p \in L = \{1, 2, \ldots, l\}$. Then, it can see that the bigger the Cor^{*,k_p} value is, the higher consistency (correlation) between the individual decision makers and the ideal decision maker. That also means that the bigger the Cor^{*,k_p} value is, the higher level of judgment the individual decision makers are on.

Suppose that $\boldsymbol{M}^* = (m_{ij}^*)_{n\times n} = (r_{ij}^*, t_{ij}^*, \pi_{ij}^*)_{n\times n}$ is such an ideal matrix that the entries in $\boldsymbol{M}^*$ are highly consistent with those in all the individual IFPRs. That is, we need to find such a matrix $\boldsymbol{M}^* = (m_{ij}^*)_{n\times n} = (r_{ij}^*, t_{ij}^*, \pi_{ij}^*)_{n\times n}$ that the following equation attains its maximum value,

$$\frac{1}{l}\sum_{p=1}^{l} C^2(m_{ij}, m_{ij}^{k_p}) = \frac{1}{l}\sum_{p=1}^{l}\frac{(r_{ij}r_{ij}^{k_p} + t_{ij}t_{ij}^{k_p} + \pi_{ij}\pi_{ij}^{k_p})^2}{(r_{ij}^2 + t_{ij}^2 + \pi_{ij}^2)({r_{ij}^{k_p}}^2 + {t_{ij}^{k_p}}^2 + {\pi_{ij}^{k_p}}^2)}, i, j \in N \tag{8.125}$$

where $\boldsymbol{M} = (m_{ij})_{n\times n} = (r_{ij}, t_{ij}, \pi_{ij})_{n\times n} \in M_n$ is an IFPR.

If we regard $(r_{ij}^{k_p}, t_{ij}^{k_p}, \pi_{ij}^{k_p})$, $(r_{ij}^*, t_{ij}^*, \pi_{ij}^*)$ and $(r_{ij}, t_{ij}, \pi_{ij})$ as vectors, we can unitize these vectors as follows:

$$(r_{ij}^{'k_p}, t_{ij}^{'k_p}, \pi_{ij}^{'k_p}) = (r_{ij}^{k_p}, t_{ij}^{k_p}, \pi_{ij}^{k_p}) / [(r_{ij}^{k_p})^2 + (t_{ij}^{k_p})^2 + (\pi_{ij}^{k_p})^2]^{\frac{1}{2}},$$

$$(r_{ij}^{'}, t_{ij}^{'}, \pi_{ij}^{'}) = (r_{ij}, t_{ij}, \pi_{ij}) / [(r_{ij})^2 + (t_{ij})^2 + (\pi_{ij})^2]^{\frac{1}{2}},$$

and

$$(r_{ij}^{*'}, t_{ij}^{*'}, \pi_{ij}^{*'}) = (r_{ij}^*, t_{ij}^*, \pi_{ij}^*) / [(r_{ij}^*)^2 + (t_{ij}^*)^2 + (\pi_{ij}^*)^2]^{\frac{1}{2}}\ (r_{ij}^*, t_{ij}^*, \pi_{ij}^*).$$

We call the following

$$\boldsymbol{M}^{'k_p} = (r_{ij}^{'k_p}, t_{ij}^{'k_p}, \pi_{ij}^{'k_p})_{n\times n},\quad p \in L, \boldsymbol{M}^{'} = (r_{ij}^{'}, t_{ij}^{'}, \pi_{ij}^{'})_{n\times n}, \boldsymbol{M}^{*'} = (r_{ij}^{*'}, t_{ij}^{*'}, \pi_{ij}^{*'})_{n\times n}$$

respectively the unitization matrices of $M^{k_p} = (r_{ij}^{k_p}, t_{ij}^{k_p}, \pi_{ij}^{k_p})_{n\times n}$, $\boldsymbol{M}^* = (m_{ij}^*)_{n\times n} = (r_{ij}^*, t_{ij}^*, \pi_{ij}^*)_{n\times n}$ and $\boldsymbol{M} = (m_{ij})_{n\times n} = (r_{ij}, t_{ij}, \pi_{ij})_{n\times n}$. Then, Eq. (8.125) is transformed into

$$\frac{1}{l}\sum_{p=1}^{l} C^2(m_{ij}, m_{ij}^{k_p}) = \frac{1}{l}\sum_{p=1}^{l} C^2(m_{ij}^{'}, m_{ij}^{'\,k_p}) = \frac{1}{l}\sum_{p=1}^{l} (r_{ij}^{'} r_{ij}^{'\,k_p} + t_{ij}^{'} t_{ij}^{'\,k_p} + \pi_{ij}^{'} \pi_{ij}^{'\,k_p})^2, i, j \in N \quad (8.126)$$

For the sake of simplicity, we also respectively regard $b = (r_{ij}^{'}, t_{ij}^{'}, \pi_{ij}^{'})^T$ and $x_p = (r_{ij}^{'\,k_p}, t_{ij}^{'\,k_p}, \pi_{ij}^{'\,k_p})^T, p \in L$, as column vectors. Then the entry $b^* = (r_{ij}^{*'}, t_{ij}^{*'}, \pi_{ij}^{*'})^T$, of $\boldsymbol{M}^*$ satisfies

$$max\frac{1}{l}\sum_{p=1}^{l} (b^T x_p)^2 = \frac{1}{l}\sum_{p=1}^{l} (b^{*T} x_p)^2 \quad (8.127)$$

Theorem 8.17. For any given

$$b = (b_1, b_2, b_3)^T = (r_{ij}^{'}, t_{ij}^{'}, \pi_{ij}^{'})^T \in R^3, \boldsymbol{X} = \begin{pmatrix} r_{ij}^{'\,k_1} & t_{ij}^{'\,k_1} & \pi_{ij}^{'\,k_1} \\ r_{ij}^{'\,k_2} & t_{ij}^{'\,k_2} & \pi_{ij}^{'\,k_2} \\ \cdots & \cdots & \cdots \\ r_{ij}^{'\,k_l} & t_{ij}^{'\,k_l} & \pi_{ij}^{'\,k_l} \end{pmatrix},$$

$max\sum_{p=1}^{l} (b^T x_p)^2 = \sum_{p=1}^{l} (b^{*T} x_p)^2 = \lambda_{max}$ holds true, where λ_{max} is the maximum eigenvalue of the matrix $\boldsymbol{F}=\boldsymbol{X}^T \boldsymbol{X}$, b^* the unique eigenvector of $\boldsymbol{F}$ that corresponds to the eigenvalue λ_{max}, satisfying $\| b^* \|_2 = 1$.

Proof. The details are similar to those of Theorem 8.15. So, they are omitted. QED.

The optimization solution $b^{*T} = (b_1^*, b_2^*, b_3^*) = (r_{ij}^{*'}, t_{ij}^{*'}, \pi_{ij}^{*'})$ of Eq. (8.127) is actually the (i,j) entry ideal matrix $\boldsymbol{M}^* = (m_{ij}^*)_{n\times n} = (r_{ij}^*, t_{ij}^*, \pi_{ij}^*)_{n\times n}$, $i, j \in N$.

According to Theorem 8.17, if $b^* = (b_1^*, b_2^*, b_3^*)^T = (r_{ij}^{*'}, t_{ij}^{*'}, \pi_{ij}^{*'})^T$ has been obtained, then by using the equations $r_{ij}^* + t_{ij}^* + \pi_{ij}^* = 1$ and $(r_{ij}^{*'}, t_{ij}^{*'}, \pi_{ij}^{*'}) = \frac{1}{(r_{ij}^{*2} + t_{ij}^{*2} + \pi_{ij}^{*2})^{\frac{1}{2}}} (r_{ij}^*, t_{ij}^*, \pi_{ij}^*)$, the value of $m_{ij}^* = (r_{ij}^*, t_{ij}^*, \pi_{ij}^*), i, j \in N$, can also be readily produced.

The next result represents a very important conclusion established in this section.

Theorem 8.18. In the set of all matrices, the ideal matrix as obtained by using Model (8.127), is such a matrix that is most consistent with all the individual IFPRs.

Proof. Let the ideal matrix $M^* = (m_{ij}^*)_{n\times n} = (r_{ij}^*, t_{ij}^*, \pi_{ij}^*)_{n\times n}$, as obtained by applying Model (8.127), be that as presented by the ideal decision maker $*$. We denote the set of all the IFPRs as $\{M_k = (m_{ij}^k)_{n\times n}, k = 1, 2, \cdots,\}$. Obviously, because $Cor^{*k_p}(i, j) \geq Cor^{kk_p}(i, j)$, we have

$$\sum_{i=1}^{n-1}\sum_{j=i+1}^{n} Cor^{*k_p}(i, j) \geq \sum_{i=1}^{n-1}\sum_{j=i+1}^{n} Cor^{kk_p}(i, j)$$

That is, $Cor^{*k_p} \geq Cor^{k,k_p}$ holds true for all $p \in L, k = 1, 2, \cdots$. QED.

8.10.2 A Numerical Example

Example 8.13. Suppose that there are three experts k_1, k_2, and k_3 who individually review three research projects s_1, s_2, and s_3. The consequent IFPRs are constructed as follows:

$$M^{k_1} = \begin{pmatrix} (1,0,0) & (0.2,0.6,0.2) & (0.4,0.5,0.1) \\ (0.6,0.2,0.2) & (1,0,0) & (0.6,0.3,0.1) \\ (0.5,0.4,0.1) & (0.3,0.6,0.1) & (1,0,0) \end{pmatrix}$$

$$M^{k_2} = \begin{pmatrix} (1,0,0) & (0.3,0.6,0.1) & (0.6,0.1,0.3) \\ (0.6,0.3,0.1) & (1,0,0) & (0.5,0.3,0.2) \\ (0.1,0.6,0.3) & (0.3,0.5,0.2) & (1,0,0) \end{pmatrix}$$

$$M^{k_3} = \begin{pmatrix} (1,0,0) & (0.2,0.6,0.2) & (0.7,0.2,0.1) \\ (0.6,0.2,0.2) & (1,0,0) & (0.6,0.4,0) \\ (0.2,0.7,0.1) & (0.4,0.6,0) & (1,0,0) \end{pmatrix}$$

According to Theorem 8.17, we need to solve for the ideal matrix $M = (m_{ij}^*)_{3\times 3}$. For the purpose of illustration, let us take the entry $m_{13}{}^*$ as an example.

Step 1: By making use of the intuitionistic fuzzy values $m_{13}^{k_1}$ = (0.4,0.5,0.1), $m_{13}^{k_2}$ = (0.6,0.1,0.3), and $m_{13}^{k_3} = (0.7, 0.2, 0.1)$, we unitize the vectors $((0.6172, 0.7715, 0.1543)^T$ $(0.8847, 0.1474, 0.4423)^T$, and $(0.9526, 0.2722, 0.1361))^T$, respectively.

Step 2: Let

$$X = \begin{pmatrix} 0.6172 & 0.7715 & 0.1543 \\ 0.8847 & 0.1474 & 0.4423 \\ 0.9526 & 0.2722 & 0.1361 \end{pmatrix}$$

Then we have

$$F = X^T X = \begin{pmatrix} 2.0711 & 0.8659 & 0.6162 \\ 0.8659 & 0.6910 & 0.2213 \\ 0.6162 & 0.2213 & 0.2380 \end{pmatrix}$$

Step 3: The maximum eigenvalue of F is 2.6632, and the corresponding unique eigenvector of F is $(0.8732, 0.4125, 0.2595)^T$. Thus we have $m_{13}^* = (0.5651, 0.2670, 0.1679)$. Similarly, we obtain that $m_{12}^* = (0.2337, 0.6000, 0.1663)$ and $m_{23}^* = (0.5696, 0.3357, 0.0946)$.

Step 4: The ultimate ideal matrix is obtained as follows:

$$M = \begin{pmatrix} (1,0,0) & (0.2337, 0.6000, 0.1663) & (0.5651, 0.2670, 0.1679) \\ (0.6000, 0.2337, 0.1663) & (1,0,0) & (0.5696, 0.3357, 0.0946) \\ (0.2670, 0.5651, 0.1679) & (0.3357, 0.5696, 0.0946) & (1,0,0) \end{pmatrix}$$

In the following, we analyze the level of judgment for each decision maker. By applying Eqs. (8.122), (8.123), and (8.124), we have

$$Cor^{*1}(1,2) = 0.9974, Cor^{*2}(1,2) = 0.9904, Cor^{*3}(1,2) = 0.9974;$$
$$Cor^{*1}(1,3) = 0.8973, Cor^{*2}(1,3) = 0.9481, Cor^{*3}(1,3) = 0.9794;$$
$$Cor^{*1}(2,3) = 0.9977, Cor^{*2}(2,3) = 0.9823, Cor^{*3}(2,3) = 0.9884.$$
$$Cor(1,2) = 0.9951, Cor(1,3) = 0.9416, Cor(2,3) = 0.9895;$$
$$Cor^{*1} = 0.9641, Cor^{*2} = 0.9736, Cor^{*3} = 0.9884.$$

These computational results show that: 1) The level of judgment of decision maker 3 is the highest, and that of decision maker 1 is the lowest. 2) The pairwise comparison between the decision alternatives s_1 and s_3 is inferior to the pairwise comparison between any other two alternatives.

8.11 Conclusion

In comparison, the concept of IFPRs can better reflect the uncertainty of behaviors and thinking of man when dealing with complex and uncertain decision making problems. However, due to the fact that the expression of IFPRs seems to be too complicated for anyone to deal with effectively, we split the expression into the interval judgment matrix of membership degrees and the interval judgment matrix of non-membership degrees.

The transitivity assumption can be used to check the consistency in the judgments of the decision makers, indicating the importance of the research into the consistency properties of preference relations. The conditions of additive and multiplicative consistencies of IFPRs are proposed on the basis of the relevant conditions established earlier for IFNPRs. As a consequence, we established the concepts of additive consistency, multiplicative consistency, general transitivity, weak transitivity, restricted max-max transitivity for IFPRs. Then, we shown that each additively (respectively, multiplicatively) consistent IFPR must also have the properties of general transitivity, weak transitivity, and restricted max-max transitivity. This fact indicates that additive (respectively, multiplicative) consistency is an elementary property of IFPRs. In many cases of decision making, the condition of additive (respectively, multiplicative) consistency is considered to be too perfect or ideal to allow for the vagueness that is inherently a part of human thinking and the complexity that naturally exists in objective things, particularly in intuitionistic circumstances. The condition of weak transitivity is a minimal requirement for making rational choices in each decision making process. Therefore, three determination theorems and corresponding algorithms were developed to judge whether or not a given IFPR has the property of weak transitivity. These algorithms can also help to produce the ranking of the decision alternatives based on IFPRs.

In real-life decision making situations, conditions of additive and multiplicative consistency are hard to satisfy. So, we turned to finding priority vectors of IFPRs. To accomplish this end, some optimization models, such as the least squares model and the goal programming model, were put forward.

The degree of membership, the degree of non-membership, and the hesitation margin of intuitionistic fuzzy sets correspond respectively to the position, negation and uncertainty of the attitude to the evidence or opinion that people holds in behavior models. So the concept of intuitionistic fuzzy sets can reflect the cognitive processes of human behaviors that appear in group decision-making situations. Therefore, the research on group decision-making based on intuitionistic fuzzy sets is more significant than that of the commonly used fuzzy sets. In this chapter, we preliminarily studied the problems of group decision making based on IFPRs:

1) We proposed approaches to intuitionistic fuzzy group decision making from three different preference points of view. And
2) We constructed consensus measures for the group IFPRs.

As the last part of this book, we use Figure 8.6 to show the relationship between different kinds of preference relations, and to summarize what is established and presented in this book.

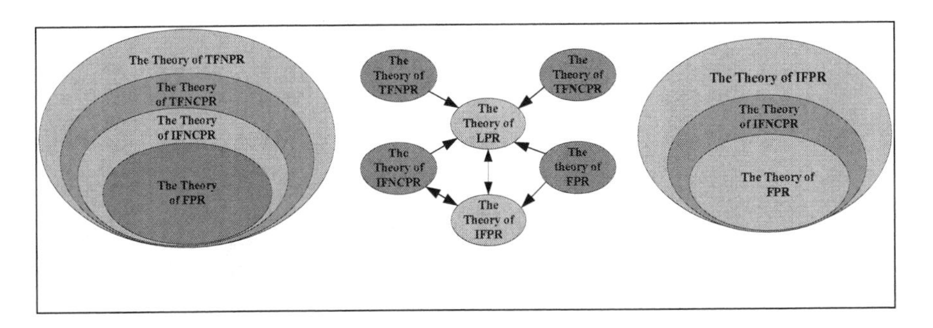

Fig. 8.6. Relationship between different kinds of preference relations

References

Ahn, B.S., Park, K.S., Han, C.H., et al.: Multi-attribute decision aid under incomplete information and hierarchical structure. European Journal of Operational Research 125(2), 431–439 (2000)

Arrow, K.J.: Social choice and individual values. Wiley, New York (1951)

Arrow, K.J.: Social choice and individual values, 2nd edn. Wiley, New York (1963)

Atanassov, K.: Intuitionistic fuzzy sets. Fuzzy Sets and Systems 20, 87–96 (1986)

Atanassov, K., Gargov, G.: Interval valued intuitionistic fuzzy sets. Fuzzy Sets and Systems 31, 343–349 (1989)

Atanassov, K.: Intuitionistic fuzzy sets: Theory and applications. Physica-Verlag, Heidelberg (1999)

Baets, B.D., Fodor, J.C.: Twenty years of fuzzy preference structures (1978–1997). Decisions in Economics and Finance 20, 45–66 (1997)

Baets, B.D., Meyer, H.D.: Transitivity frameworks for reciprocal relations: Cycle-transitivity versus FG-transitivity. Fuzzy Sets and Systems 152, 249–270 (2005)

Baets, B.D., Meyer, H.D.: Transitive approximation of fuzzy relations by alternating closures and openings. Soft Computing- A Fusion of Foundations, Methodologies and Applications 7, 210–219 (2008)

Baets, B.D., Meyer, H.D., Schuymer, B.D., Jenei, S.: Cyclic evaluation of transitivity of reciprocal relations. Social Choice and Welfare 26, 217–238 (2006)

Baets, B.D., Walle, B.V., Kerre, E.: Fuzzy preference structures without incomparability. Fuzzy Sets and Systems 76, 333–348 (1995)

Balopoulos, V., Hatzimichailidis, A.G., Papadopoulos, B.K.: Distance and similarity measures for fuzzy operators. Information Sciences 177, 2336–2348 (2007)

Bisdorff, R., Roubens, M.: Choice Procedures in Pairwise Comparison Multiple-Attribute Decision Making Methods. In: Berghammer, R., Möller, B., Struth, G. (eds.) RelMiCS 2003. LNCS, vol. 3051, pp. 1–7. Springer, Heidelberg (2004)

Buckley, J.J.: Fuzzy hierarchical analysis. Fuzzy Sets and Systems 17, 233–247 (1985)

Bustince, H., Burillo, P.: Vague sets are intuitionistic fuzzy sets. Fuzzy Sets and Systems 79, 403–405 (1996)

Carmone, F.J., Kara, A., Zanakis, S.H.: A Monte Carlo investigation of incomplete pairwise comparison matrices in AHP. European Journal of Operational Research 102(3), 538–553 (1997)

Chakraborty, M.C., Chakraborty, D.: A theoretical development on fuzzy distance measure for fuzzy numbers. Mathematical and Computer Modelling 43, 254–261 (2006)

Chang, D.Y.: Applications of the extent analysis method on fuzzy AHP. European Journal of Operational Research 95, 649–655 (1996)

Chen, S.M., Chen, J.H.: Fuzzy risk analysis based on similarity measures between interval-valued fuzzy numbers and interval-valued fuzzy number arithmetic operators. Expert Systems with Applications 36, 6309–6317 (2009)

Chen, X., Fan, Z.P.: Research on quality of evaluating experts using matrices of their language evaluations. Systems Engineering 24(4), 111–115 (2006)

Chen, Y., Fan, Z.P.: Consistency and relevant problems of judgment matrices of language. Systems Engineering: Theory and Practice 24(1), 136–151 (2004)

Chiclana, F., Herrera, F., Herrera-Viedma, E., et al.: A note on the reciprocity in the aggregation of fuzzy preference relations using OWA operations. Fuzzy Sets and Systems 137, 71–83 (2003)

Delgado, M., Herrera, F., Herrera, E.: A communication model based on the 2-tuple fuzzy linguistic representation for distributed intelligent agent system on internet. Soft Computing 6, 320–328 (2002)

Deng, J.L.: Grey group decision in grey relational space. The Journal of Grey System (3), 177–182 (1998)

Díaz, S., Baets, B.D., Montes, S.: General results on the decomposition of transitive fuzzy relations. Fuzzy Optim. Decis. Making 9, 1–29 (2010)

Díaz, S., Montes, S., Baets, B.D.: Transitive decomposition of fuzzy preference relations: The case of nilpotent minimum. Kybernetika 40, 71–88 (2004)

Dimitrov, D.: The paretian liberal with intuitionistic fuzzy preferences: a result. Social Choice Welfare 23, 149–156 (2004)

Dubois, D., Gottwald, S.: Terminological difficulties in fuzzy set theory-The case of "Intuitionistic Fuzzy Sets". Fuzzy Sets and Systems 156, 485–491 (2005)

Fan, Z.P., Jiang, Y.P.: A survey on the ranking research of fuzzy judgment matrices. Systems Engineering 19(5), 12–18 (2001)

Fan, Z.P., Jiang, Y.P.: Satisfactory consistency of judgment matrices of language. Control and Decision Making 19(8), 903–906 (2004)

Filev, D., Yager, R.R.: Analytic properties of maximum entropy OWA operator. Information Sciences 85, 11–27 (1995)

Fishburn, P.C.: The Theory of Social Choice. Princeton University Press, Princeton (1973)

Fodor, J.C., Roubens, M.: Valued preference structures. European Journal of Operational Research 79, 277–286 (1994)

Fodor, J.C.: An axiomatic approach to fuzzy preference modeling. Fuzzy Sets and Systems 52, 47–52 (1992)

Fodor, J.C., Orlovski, S.: The use of fuzzy preference models. In: Perny, P., Roubens, M. (eds.) Multiple Criteria Choice, Ranking and Sorting. Fuzzy Sets in Decision Analysis, Operations Research and Statistics. The Kluwer Handbook Series on Fuzzy Sets, vol. 1, pp. 69–101 (1998)

Freson, S., De Meyer, H., De Baets, B.: An Algorithm for Generating Consistent and Transitive Approximations of Reciprocal Preference Relations. In: Hüllermeier, E., Kruse, R., Hoffmann, F. (eds.) Computational Intelligence for Knowledge-Based Systems Design (IPMU 2010). LNCS, vol. 6178, pp. 564–573. Springer, Heidelberg (2010)

Gao, L.S.: The fuzzy arithmetic mean. Fuzzy Sets and Systems 107(1), 335–348 (1999)

Gisella, F., Roberto, G.R.: A characterization of a general class of ranking functions on triangular fuzzy numbers. Fuzzy Sets and Systems 146(2), 297–312 (2004)

Gogus, O., Boucher, T.O.: Strong transitivity rationality and weak monotonicity in fuzzy pairwise comparisons. Fuzzy Sets and Systems 94(1), 133–144 (1998)

Gong, Z.W.: Theory and Methods of Uncertain Fuzzy Judgment Matrices. PhD Dissertation of Nanjing University of Aeronautics and Astronautics, Nanjing, China (2006)

Gong, Z.W.: On ranking method of complementary judgment matrices of trapezoidal fuzzy numbers based on binary semantics. Systems Engineering and Electronic Technology 29(9), 1488–1492 (2007)

Gong, Z.W.: Least square method to priority of the fuzzy preference relation with incomplete information. International Journal of Approximate Reasoning 47(2), 258–264 (2008)

Gong, Z.W., Bai, M.G.: Ranking method of fuzzy judgment matrices using relative entropy. In: Proceedings of the 6th Chinese Annual Conference of Uncertain Systems, Luoyang, pp. 185–191 (2008)

Gong, Z.W., Bai, M.G., Sun, R.L.: The logarithmic least squares priority model of the collective IFNPRs. In: Chinese Control and Decision Conference, Guilin, pp. 5954–5957 (2009)

Gong, Z.W., Cui, W.J.: Approach to group decision making based on the incomplete interval fuzzy number complementary judgment matrices. In: IEEE International Conference on Fuzzy Systems, Hong Kong, pp. 650–654 (2008)

Gong, Z.W., Li, L.S., Cao, J., et al.: On additive consistent properties of the intuitionistic fuzzy preference relation. International Journal of Information Technology & Decision Making 9(6), 1009–1025 (2010)

Gong, Z.W., Li, L.S., Cao, J., Zhao, Y.: The optimal priority models of the intuitionistic fuzzy preference relation and their application in selecting industries with higher meteorological sensitivity. Expert Systems with Applications 38, 4394–4402 (2011)

Gong, Z.W., Li, L.S., Yao, T.X.: Group decision making based on incomplete intuitionistic fuzzy preference relations. Operations Research and Management Science 19(4), 45–51 (2010)

Gong, Z.W., Li, L.S., Zhou, F.X., et al.: Goal programming approaches to obtain the priority vectors from the intuitionistic fuzzy preference relations. Computers & Industrial Engineering 57, 1187–1193 (2009)

Gong, Z.W., Liu, S.F.: Consistency and ranking method of complementary preference relations of interval numbers. Management Science of China 14(4), 64–68 (2006a)

Gong, Z.W., Liu, S.F.: Properties of relevant problems of complementary judgment matrices of interval numbers. Operations Research and Management 15(6), 25–30 (2006b)

Gong, Z., Liu, S.: On Properties and the Corresponding Problems of Triangular Fuzzy Number Complementary Preference Relations. In: Wang, L., Jiao, L., Shi, G., Li, X., Liu, J. (eds.) FSKD 2006. LNCS (LNAI), vol. 4223, pp. 334–343. Springer, Heidelberg (2006)

Gong, Z.W., Liu, S.F.: Consistency and ranking of complementary judgment matrices of triangular fuzzy numbers. Control and Decision Making 21(8), 903–907 (2006d)

Gong, Z.W., Liu, S.F.: The ranking of reciprocal judgment matrices with minimal Theil unequal coefficients. Journal of Guangxi University 31(1), 49–53 (2006e)

Gong, Z.W., Liu, S.F.: A binary linguistic decision method based on group judgment matrices of different fuzzy preferences. Journal of Systems Engineering 22(2), 185–189 (2007a)

Gong, Z.W., Liu, S.F.: Inverse problems in group decision making using intuitionistic fuzzy judgment matrices. Journal of Systems Management 16(5), 497–501 (2007b)

Gong, Z.W., Liu, S.F.: Consistency and relevant problems of binary linguistic judgment matrices. Journal of Nanjing University of Aeronautics and Astronautics 39(4), 550–554 (2007c)

Gong, Z.W., Wu, J., Cui, W.J.: Group decision making methods for different fuzzy preferences with uncertainty number. In: Proceedings of the 5th International Conference on Fuzzy Systems and Knowledge Discovery, FSKD 2008, Jinan, pp. 296–300 (2008)

Gong, Z.W., Yao, T.X.: On the priority models of the grey interval preference relation. In: IEEE International Conference on Grey Systems and Intelligent Service, Nanjing (2009)

Gong, Z.W., Liu, S.F.: Engenvector ranking of fuzzy judgment matrices. Operations Research and Management 15(4), 27–31 (2006)

Gong, Z.W., Zhang, L.F., Liu, S.F.: Group decision making of complementary preferences of triangular fuzzy numbers under remnant information. Journal of Systems Engineering 23(3), 269–275 (2008)

Guha, D., Chakraborty, D.: A new approach to fuzzy distance measure and similarity measure between two generalized fuzzy numbers. Applied Soft Computing 10, 90–99 (2010)

Guo, C.X., Guo, Y.H.: Group decision making with different attribute forms of preference. Systems Engineering and Electronic Technology 27(1), 63–65 (2005)

Hammond, K.R.: Human Judgment and Social Policy: Irreducible Uncertainty, Inevitable Error, Unavailable Injustice. Oxford University Press, New York (1996)

Herrera, E.: An information retrieval model with ordinal linguistic weighed queries based on two weighting elements information in group decision making. International Journal of Uncertainty Fuzziness and Knowledge-Based Systems 9(supplementary), 77–87 (2001)

Herrera, F., Herrera, E., Martinez, L.: A fusion approach for managing multi-granularity linguistic term sets in decision making. European Journal of Operational Research 114, 43–58 (2000)

Herrera, F., Martinez, L.: An approach for combing linguistic and numerical information based on the 2-tuple fuzzy linguistic representation model in decision making. International Journal of Uncertainty Fuzziness and Knowledge-Based Systems 8(5), 539–562 (2000)

Herrera, F., Martinez, L.: The 2-tuple linguistic computational model advantages of its linguistic description accuracy and consistency. International Journal of Uncertainty Fuzziness and Knowledge-Based Systems 9(supplementary), 33–48 (2001a)

Herrera, F., Martinez, L.: A model based on linguistic 2-tuples for dealing with multi-granular hierarchical linguistic contexts in multi-expert decision-making. IEEE Transactions on Systems Man and Cybernetics-Part B: Cybernetics 31(2), 227–234 (2001b)

Herrera, F., Martinez, L., Sanchez, P.J.: Managing non-homogeneous information in group decision making. European Journal of Operational Research 166(1), 115–132 (2005)

Herrera, F., Herrera-Viedma, E., Verdegay, J.L.: A sequential selection process in group decision making with a linguistic assessment approach. Information Science 85, 223–239 (1995)

Herrera, F., Herrera-Viedma, E., Verdegay, J.L.: Direct approach processes in group decision making using linguistic OWA operators. Fuzzy Sets and Systems 79, 175–190 (1996a)

Herrera-Viedma, E., Herrera, F., Chiclana, F., et al.: Some issues on consistency of fuzzy preference relations. European Journal of Operational Research 154(1), 98–109 (2004)

Horn, R.A., Johnson, C.R.: Matrix Analysis. Cambridge University Press, Cambridge (1990)

Hsu, H.M., Chen, C.T.: Aggregation of fuzzy opinions under group decision making. Fuzzy Sets and Systems 79, 279–285 (1996)

Hu, Y.C., Tsai, J.F.: Backpropagation multi-layer perception for incomplete pairwise comparison matrices in analytic hierarchy process. Applied Mathematics and Computation 180(1), 53–62 (2006)

Hung, W.L., Yang, M.S.: Similarity measures of intuitionistic fuzzy sets based on Hausdorff distance. Pattern Recognition Letters 25, 1603–1611 (2004)

Huo, F.J., Wu, Z.Z.: Consistency and ranking of complementary judgment matrices of the I-type uncertain numbers. Systems Engineering: Theory and Practice 25(10), 60–66 (2005a)

Huo, F.J., Wu, Q.Z.: The ranking of complementary judgment matrices of fuzzy numbers of Hausdorff distance. Fuzzy Systems and Mathematics 19(2), 110–115 (2005b)

Hyde, K.M., Maier, H.R., Colby, C.B.: A distance-based uncertainty analysis approach to multi-criteria decision analysis for water resource decision making. Journal of Environmental Management 77, 278–290 (2005)

Jiang, Y.P., Fan, Z.P.: A ranking method of complementary judgment matrices of one type triangular fuzzy numbers. Systems Engineering and Electronic Technology 24(7), 34–36 (2002a)

Jiang, Y.P., Fan, Z.P.: A practically useful ranking method of complementary judgment matrices of triangular fuzzy numbers. Systems Engineering 20(2), 89–92 (2002b)

Jiang, Y.P., Fan, Z.P.: Properties of aggregation operators of binary linguistic information. Control and Decision Making 18(6), 754–757 (2003)

Jiang, Y.P., Fan, Z.S.: Theory and Methods of Decision Making Based on Judgment Matrices. Science Press, Beijing (2008)

Jose, L.G.: A general class of simple majority decision rules based on linguistic opinions. Information Science 176(1), 352–365 (2006)

Kim, S.H., Ahn, B.S.: Group decision making procedure considering preference strength under incomplete information. Computers Ops. Res. 24(12), 1101–1112 (1997)

Kim, S.H., Ahn, B.S.: Interactive group decision making procedure under incomplete information. European Journal of Operational Research 116(3), 498–507 (1999)

Kim, S.H., Choi, S.H., Kim, J.: An interactive procedure for multiple criteria group decision making with incomplete information. Computers and Industrial Engineering 35(1), 295–298 (1998)

Kim, S.H., Choi, S.H., Kim, J.: An interactive procedure for multiple attribute group decision making with incomplete information: Range-based approach. European Journal of Operational Research 118(1), 139–152 (1999)

Kim, S.H., Han, C.H.: An interactive procedure for multi-attribute group decision making with incomplete information. Computers & Operations Research 118(1), 139–152 (1999)

Kwiesielewicz, M.: A note on the fuzzy extension of Saaty's priority theory. Fuzzy Sets and Systems 953, 161–172 (1998)

Lee, H.S.: Optimal consensus of fuzzy opinions under group decision making environment. Fuzzy Sets and Systems 132, 303–315 (2002)

Lee, S.M.: Goal programming for decision analysis. Auerbach Publishing Co., Philadelphia (1973)

Lei, Y.J., Wang, B.S., Miao, G.Q.: Intuitionistic fuzzy relations and their composite operations. Systems Engineering: Theory and Practice 25(2), 113–118 (2005)

Li, B.J., Liu, S.F.: A new group decision making method based on judgment matrices of a kind of interval numbers. Management Science of China 12(6), 109–112 (2004)

Li, D.F.: Multiattribute decision making models and methods using intuitionistic fuzzy sets. Journal of Computer and System Sciences 70(1), 73–85 (2005)

Li, D.F., Cheng, C.T.: New similarity measures of intuitionistic fuzzy sets and application to pattern recognition. Pattern Recognition Letters 23, 221–225 (2002)

Li, D.F., Wang, Y.C., Liu, S.: Fractional programming methodology for multi-attribute group decision-making using IFS. Applied Soft Computing 9, 219–225 (2008)

Li, H.L., Ma, L.C.: Visualizing decision process on spheres based on the even swap concept. Decision Support Systems 45, 354–367 (2008)

Li, Q.: Eight Lectures on Matrix Theory. Shanghai Press of Science and Technology, Shanghai (1988)

Li, Y.H., Olson, L.D., Qin, Z.: Similarity measures between intuitionistic fuzzy (vague) sets: A comparative analysis. Pattern Recognition Letters 28, 278–285 (2007)

Lin, L., Yuan, X.H., Xia, Z.Q.: Multicriteria fuzzy decision-making methods based on intuitionistic fuzzy sets. J. Comput. System Sci. 73, 84–88 (2007)

Liu, H.W., Wang, G.J.: Multi-criteria decision-making methods based on intuitionistic fuzzy sets. European Journal of Operational Research 179, 220–233 (2007)

Liu, S.F., Lin, Y.: Grey Information: Theory and Practical Applications. Springer, New York (2006)

Ma, L.C.: Visualizing preferences on spheres for group decisions based on multiplicative preference relations. European Journal of Operational Research 203, 176–184 (2010)

Moore, R.E.: Method and Application of Interval Analysis. Prentice Hall, London (1979)

Moore, R., Lodwick, W.: Interval analysis and fuzzy set theory. Fuzzy Sets and Systems 135, 5–9 (2003)

Nishizawa, K.: A method to find elements of cycles in an incomplete directed graph and its applications binary AHP and Petri nets. Computers Math. Applic. 33(9), 33–46 (1997)

Orlovsky, S.A.: Decision making with a fuzzy preference relation. Fuzzy Sets and Systems 3(1), 155–167 (1978)

Ovchinnikov, S., Roubens, M.: On fuzzy strict preference, indifference, and incomparability relations. Fuzzy Sets and Systems 49, 15–20 (1992)

Qiu, Y.H.: Management Decision Analysis and applied Entropy. Press of Machinery Industry, Beijing (2002)

Pankowska, A., Wygralak, M.: General IF-sets with triangular norms and their applications to group decision making. Information Sciences 176, 2713–2754 (2006)

Pasi, G., Yager, R.R.: Modeling the concept of majority opinion in group decision making. Information Sciences 176(4), 390–414 (2006)

Rademaker, M., Baets, B.D.: Consistent union and prioritized consistent union: New operations for preference aggregation. Ann. Oper. Res. (2011) (published online)

Saaty, T.L.: The Analytic Hierarchy Process. McGraw-Hill, New York (1980)

Sen, A.K.: Preferences, votes and the transitivity of majority decisions. Review of Economic Studies 31(2), 163–165 (1964)

Sen, A.K.: Collective Choice and Social Welfare. Holden-Day, San Francisco (1970)

Song, G.X., Yang, D.L.: Mutual Transformations between AHP and fuzzy judgment matrices. Journal of Daliang University of Technology 23(4), 535–539 (2003)

Sugihara, K., Ishii, H., Tanaka, H.: Interval priorities in AHP by interval regression analysis. European Journal of Operational Research 158, 745–754 (2004)

Supriya, K.D., Ranjit, B., Akhil, R.R.: An application of intuitionistic fuzzy sets in medical diagnosis. Fuzzy Sets and Systems 117(2), 209–213 (2001)

Szmidt, E., Kacprzyk, J.: Distance between intuitionistic fuzzy sets. Fuzzy Sets Systems 114(3), 505–518 (2000)

Szmidt, E., Kacprzyk, J.: A consensus-reaching process under intuitionistic fuzzy preference relations. International Journal of Intelligent Systems 18, 837–852 (2003)

Szmidt, E., Kacprzyk, J.: Similarity of intuitionistic fuzzy sets and the jaccard coefficient. In: IPMU 2004 Proceedings, Perugia, Italy, pp. 1405–1412 (2004)

Szmidt, E., Kacprzyk, J.: A New Concept of a Similarity Measure for Intuitionistic Fuzzy Sets and Its Use in Group Decision Making. In: Torra, V., Narukawa, Y., Miyamoto, S. (eds.) MDAI 2005. LNCS (LNAI), vol. 3558, pp. 272–282. Springer, Heidelberg (2005)

Szmidt, E., Kacprzyk, J.: Distances between intuitionistic fuzzy sets and their applications in reasoning. SCI, vol. 2, pp. 101–116. Springer, Heidelberg (2005b)

Takeda, E.: A method for multiple pseudo-criteria decision problems. Computers & Operations Research 28(13), 1427–1439 (2001)

Tan, C.Q., Zhang, Q.: Clustering analysis of experts' opinions using intuitionistic fuzzy distances. Practice and Comprehension of Mathematics 36(2), 119–124 (2006)

Tanino, T.: Fuzzy preference orderings in group decision making. Fuzzy Sets and Systems 12, 117–131 (1984)

Tizhoosh, H.R.: Interval-valued versus intuitionistic fuzzy sets: Isomorphism versus semantics. Pattern Recognition 41, 1812–1813 (2008)

Tang, X.L., Gong, Z.W.: Decision making methods using intuitionistic fuzzy matrices. Systems Engineering 25(6), 79–83 (2007)

Ullah, A.: Entropy divergence and distance measures with econometric applications. Journal of Statistical Planning and Inference 49, 137–162 (1996)

Van Laarhoven, P.J.M., Pedrcyz, W.: A fuzzy extension of Saaty's priority theory. Fuzzy Sets and Systems 11, 229–241 (1983)

Viglioccoa, G., Vinsona, D.P., Damianb, M.F., et al.: Semantic distance effects on object and action naming. Cognition 85, 61–69 (2002)

Vlachos, I.K., Sergiadis, G.D.: Intuitionistic fuzzy information-Applications to pattern recognition. Pattern Recognition Letters 28, 197–206 (2005)

Wan, Y.C., Sheng, Z.Y.: Hierarchical analysis method using unknown triple-valued judgments. Systems Engineering: Theory and Practice 24(12), 89–93 (2004)

Wang, L.F., Xu, L.F.: Introduction to Hierarchical Analysis. Press of People's University of China, Beijing (1990)

Wang, P.: QoS-aware web services selection with intuitionistic fuzzy set under consumer's vague perception. Expert Systems with Applications 36, 4460–4466 (2009)

Wang, R.C., Chuu, S.J.: Group decision-making using a fuzzy linguistic approach for evaluating the flexibility in a manufacturing system. European Journal of Operational Research 154(1), 563–572 (2004)

Wang, X.R., Fan, Z.P.: A linguistic decision making method based on information of binary semantics. Journal of Management Science 6(5), 1–5 (2003)

Wang, X.Z., Liu, J.S., Wei, Y.Q.: Consistency and ranking of weights pf fuzzy judgment matrices. Systems Engineering: Theory and Practice 15(1), 28–35 (1995)

Wang, Y.M.: A survey on methods of ranking judgment matrices. Decision Making and Decision Support Systems 5(3), 101–104 (1995)

Wang, Y.M.: On lexicographic goal programming method for generating weights from inconsistent interval comparison matrices. Applied Mathematics and Computation 173, 985–991 (2006)

Wang, Y.M., Chin, K.S.: A linear goal programming priority method for fuzzy analytic hierarchy process and its applications in new product screening. International Journal of Approximate Reasoning 49(2), 451–465 (2008)

Wang, Y.M., Elhag, T.M.S.: On the normalization of interval and fuzzy weights. Fuzzy Sets and Systems 157, 2456–2471 (2006)

Wang, Y.M., Elhag, T.M.S.: A goal programming method for obtaining interval weights from an interval comparison matrix. European Journal of Operational Research 177, 458–471 (2007)

Wang, Y.M., Elhag, T.M.S., Hua, Z.S.: A modified fuzzy logarithmic least squares method for fuzzy analytic hierarchy process. Fuzzy Sets and Systems 157, 3055–3071 (2006)

Wang, Y.M., Fan, Z.P.: Fuzzy preference relations: Aggregation and weight determination. Computers & Industrial Engineering 53, 163–172 (2007)

Wang, Y.M., Luo, Y., Hua, Z.S.: On the extent analysis method for fuzzy AHP and its applications. European Journal of Operational Research 186, 735–747 (2008)

Wei, Y.Q., Liu, J.S., Wang, X.Z.: Consistency and weights in uncertain AHP judgment matrices. Systems Engineering: Theory and Practice 14(7), 16–22 (1994)

Wu, J.: A method of aggregation of preference information of complementary judgment matrices of group interval numbers. Systems Engineering: Theory and Practice 13(6), 500–503 (2004)

Xiao, S.H., Fan, Z.P., Wang, M.G.: Two kinds of preference information in group decision making: A unified method of AHP judgment and fuzzy preference matrices. Journal of Systems Engineering 17(1), 82–86 (2002)

Xiao, Y., Li, H.: Improvement and application of judgment matrices of triangular fuzzy numbers. Fuzzy Systems and Mathematics 17(2), 59–64 (2003)

Xu, S.B.: Principles of Hierarchical Analysis. Press of Tianjing University, Tianjing (1988)

Xu, X.Y., Yang, Y.Q.: Consistent approximation and ranking of uncertain AHP judgment matrices. Systems Engineering: Theory and Practice 18(2), 19–22 (1998)

Xu, Z.S.: Two ranking methods of complementary judgment matrices: Least squares of weights and eigenvectors. Systems Engineering: Theory and Practice 22(7), 71–75 (2002)

Xu, Z.S.: A method of ranking complementary judgment matrices of triangular fuzzy numbers. Fuzzy Systems and Mathematics 16(1), 47–50 (2003)

Xu, Z.S.: On the ranking methods of complementary judgment matrices of triangular fuzzy numbers. Journal of Systems Engineering 19(1), 85–88 (2004a)

Xu, Z.S.: Methods of Uncertain, Multi-Attributes Decision Making. Press of Qinghua University, Beijing (2004b)

Xu, Z.S.: Incomplete complementary judgment matrices. Systems Engineering: Theory and Practice 24(6), 91–97 (2004c)

Xu, Z.S.: Goal programming models for obtaining the priority vector of incomplete fuzzy preference relation. International Journal of Approximate Reasoning 36(3), 261–270 (2004d)

Xu, Z.S.: Group decision making with pure linguistic multi-attributes. Control and Decision Making 19(7), 778–786 (2004e)

Xu, Z.S.: Group decision making based on crossed over incomplete complementary judgment matrices. Control and Decision Making 20(8), 913–916 (2005a)

Xu, Z.S.: Deviation measures of linguistic preference relations in group decision making. Omega 33, 249–254 (2005b)

Xu, Z.S.: Multi-attribute group decision making based on the terminological index of language scale. Journal of Systems Engineering 20(1), 84–88 (2005c)

Xu, Z.S.: Group decision making based on different kinds of incomplete judgment matrices. Control and Decision Making 21(1), 28–33 (2006)

Xu, Z.S.: A survey of preference relations. International Journal of General Systems 36, 179–203 (2007a)

Xu, Z.S.: Intuitionistic preference relations and their application in group decision making. Information Sciences 177, 2363–2379 (2007b)

Xu, Z.S.: Theory and Application of the Aggregation of Intuitionistic Information. Science Press, Beijing (2008)

Xu, Z.S., Da, Q.L.: A ranking method of interval number judgment matrices using degrees of possibility. Management Science of China 11(2), 63–65 (2003a)

Xu, Z.S., Da, Q.L.: Possibility ranking method and application of interval numbers. Journal of Systems Engineering 18(1), 67–70 (2003b)

Xu, Z.S., Yager, R.R.: Some geometric aggregation operators based on intuitionistic fuzzy sets. International Journal of General Systems 35, 417–433 (2006)

Yager, R.R.: Applications and extensions of OWA aggregations. International Journal of Man-Machine Studies 37, 103–132 (1992)

Yager, R.R.: An approach to ordinal decision making. International Journal of Approximate Reasoning 12, 237–261 (1995)

Yager, R.R.: OWA aggregation over a continuous interval argument with applications to decision making. IEEE Transactions on Systems Man and Cybernetics-Part B, Cybernetics 34(5), 1952–1963 (2004)

Yang, Q.: Matrix Analysis. Press of Tianjing University, Tianjing (1989)

Yang, Y., Chiclana, F.: Intuitionistic fuzzy sets: Spherical representation and distances. International Journal of Intelligent Systems 24, 399–420 (2009)

Ye, J.: Multicriteria fuzzy decision-making method based on a novel accuracy function under interval-valued intuitionistic fuzzy environment. Expert Systems with Applications 36, 6899–6902 (2009)

You, T.H., Fan, Z.P., Li, H.Y.: A method that comprehensively evaluates the software that deals with binary linguistic information. Systems Engineering and Electronic Technology 27(3), 545–549 (2005)

Young, R.C.: The algebra of many valued quantities. Annals of Mathematics 31, 260–290 (1931)

Yu, C.H., Fan, Z.P.: A method of clustering multiply indexed information based on triangular fuzzy numbers. Systems Engineering: Theory and Practice 13(5), 467–470 (2004)

Zadeh, L.A.: Fuzzy sets. Information and Control 18, 338–353 (1965)

Zadeh, L.A.: The concept of a linguistic variable and its application to approximate reasoning (Part I). Information Sciences 8, 199–249 (1975)

Zadrozny, S., Kacprzyk, J.: Computing with words for text processing: an approach to the text categorization. Information Sciences 176(4), 415–437 (2006)

Zhang, G., Lu, J.: An integrated group decision-making method dealing with fuzzy preferences for alternatives and individual judgments for selection criteria. Group Decision and Negotiation 12, 501–515 (2003)

Zhang, J.J.: On methods of ordering interval numbers. Operations Research and Management 12(3), 18–22 (2003)

Zhang, L.D., Gong, Z.W.: On the competitive environment and strategy of Chinese petroleum companies. Journal of Petroleum University of China (Social Science Edition) 22(4), 12–15 (2006)

Zhang, L.F., Li, D., Gong, Z.W.: Evaluation of sustainable development of petroleum firms using rough FAHP method. Journal of Petroleum University of China (Natural Science Edition) 30(6), 145–154 (2006)

Zhu, K.J., Jing, Y., Chang, D.Y.: A discussion on extent analysis method and applications of fuzzy AHP. European Journal of Operational Research 116(2), 450–456 (1999)

Index

Printed by Publishers' Graphics LLC